Basic Electrodynamics in 6 Lessons

Martin Poppe

Basic Electrodynamics in 6 Lessons

 Springer

Martin Poppe ⓘ
FB Elektrotechnik und Informatik
Fachhochschule Münster
Steinfurt, Nordrhein-Westfalen
Germany

ISBN 978-3-662-69142-7   ISBN 978-3-662-69143-4   (eBook)
https://doi.org/10.1007/978-3-662-69143-4

This Springer imprint is published by the registered company Springer-Verlag
GmbH, DE, part of Springer Nature.
The registered company address is: Heidelberger Platz 3, 14197 Berlin, Germany

If disposing of this product, please recycle the paper.

Preface

The language of truth is simple

—Euripides

"Finally, I got it right—but honestly, I don't know why". This is one of the most frequently heard sentences in connection with electrodynamics. The know-how is there, and the algorithms are known, but the meaning is only sort of straightforward. What about "Finally, I got it right, and I even know why", instead? I wrote this little book to promote that phrase. The six lessons are intended to provide a profound understanding of electrodynamics thus enabling the reader to depart from standard algorithms and techniques. Because only the knowledge of one's ability creates the self-confidence that a person needs, to tackle new things.

The straight path leads fastest to understanding

This book is intended to use the shortest route to a basic understanding of electrodynamics. This is possible because this theory is entirely based on a very small number of quantities:
1. a single quality of matter called charge, Q
2. the fields of the electric force E and the field of the magnetic force B and their coupling constants ε_0 and μ_0.

Their relationships are entirely determined by the Coulomb force, the Lorentz force, and Maxwell's equations. Everything is based there-upon—in electrodynamics and this book. This book tries to fulfil Albert Einstein's uncompromising demand for simplicity: "The primary goal of all theory is to make the irreducible elements as simple and as few in number as possible, without relinquishing the accurate representation of any experience."

Applications are the purpose of electrodynamics

This book only considers a topic to have been dealt with after an application could be addressed. There are examples as well as questions and problems. This allows readers to monitor their learning progress continuously because the following sentence can be read both ways: "If one understands something, one can apply it."

There will be days when a reader does not reach peak performance. If, on such a day, a problem in this book seems unsolvable, this should not be a cause for depression. Instead, it should be motivation to use the solutions as supplementary teaching material to deepen one's understanding.

What nature has joined together, let not man put asunder

A key question in modern theories is that of distinguishability. In quantum mechanics, it is precisely its negation that plays a central role. The concept of fundamentally indistinguishable parts is an indispensable prerequisite to understanding the Pauli principle, quantum statistics, and phenomena such as superconductivity.

After reading this book, the reader will realise that the indistinguishability of field components plays a similarly central role in understanding classical electrical engineering. It shines a different light on various topics: Questions such as "What does the displacement current displace?", "Why can you add a field to a dipole density, but not pears and apples?" or even "How does one find the poles in the H-field" are not answered; they simply disappear. And the correct answer to how many Maxwell equations there are comes out quite naturally: "Generally four, but six in the presence of matter".

Finally, this is the English edition of "Grundkurs Theoretische Elektrotechnik". If you happen to be the proud owner of that book, there is no need to buy this one.

Martin Poppe
Steinfurt, Germany
March 2024

Acknowledgements I want to thank Ms. Eva Hestermann and Mr. Michael Kottusch. Their support was a vital ingredient for publishing this book. I am also grateful to my colleagues Erik Baak, Andreas Boeker, Ralf Hinterding, Rainer Nix, and Ulrich Wittrock for their helpful comments.

This book could only be written because of Ulrike's continuing support.

Contents

Acronyms

A	Magnetic vector potential
$\boldsymbol{a}$	Surface normal vector
a	Surface area
$\boldsymbol{B}$	Field of magnetic force
B	Magnitude of $\boldsymbol{B}$
C	Capacitance, $Q = C \cdot U$
C_l	Capacitance per unit length
$C(x)$	Field theory: charge conjugation operation on x
c	Vacuum speed of light
$\boldsymbol{D}$	Electric dipole density
ϵ_0	Electric permittivity
ϵ_r	Relative electric permittivity
$\boldsymbol{F}$	Force, $\mathbf{F} = \mathrm{d}\boldsymbol{p}/\mathrm{d}t$
f	Frequency
$\Phi(\boldsymbol{r})$	Electric potential at location $\boldsymbol{r}$
Φ_B	Flux of field B
	Momentum density
$\boldsymbol{H}$	"Magnetic field strength"
I	Current $I = \mathrm{d}Q/\mathrm{d}t$
$\hat{I}$	Amplitude of alternating current
$\underline{I}$	Compex amplitude of alternating current
i	Harmonically alternating current
$\underline{i}$	Complex extension of alternating current $i = \mathfrak{J}(\underline{i})$
$\boldsymbol{J}$	Current density, $\boldsymbol{J} = \rho \cdot \boldsymbol{v}$
$\boldsymbol{k}$	Wave vector
L	Inductance, $U_{ind} = L\frac{\mathrm{d}I}{\mathrm{d}t}$
L_l	Inductance per unit length
λ	Wave length
$\boldsymbol{M}$	Magnetic dipole density or magnetisation
$\boldsymbol{m}$	Magnetic dipole moment
μ_0	Magnetic permeability of the vacuum
μ_r	Relative magnetic permeability
$P(\boldsymbol{r})$	Parity operation on $\boldsymbol{r}$
$\boldsymbol{P}$	Dipole density or polarisation
$\boldsymbol{p}$	Electric dipole moment
Q	Charge

R	Ohmic resistance
ρ	1. Field theory: charge density
	2. Circuits: specific resistance
S	Poynting vector
σ	Specific conductivity
T	1. Engineering: time span of one period, the inverse of frequency f
	2. Solid state physics: absolute temperature
$T(x)$	Time reversal operation on x
t	Time
τ	Torque
τ_-	Reduced time $\tau_- = t - r/v$
U	Voltage or tension
$\hat{U}$	Amplitude of alternating voltage
$\underline{U}$	Compex amplitude of alternating voltage
u	Harmonically alternating voltage
$\underline{u}$	Complex extension of alternating voltage $u = \Im(\underline{u})$
V	Volume
v	Velocity or speed, $v = ds/dt$
W	Energy
$\boldsymbol{w}$	Energy density
$\underline{Z}$	Impedance
$\underline{Z}_0$	Characteristic impedance
ω	Angular frequency, $\omega = 2\pi \cdot f$

Introduction: Classical Electrodynamics in Modern Times

Contents

© The Author(s), under exclusive license to Springer-Verlag GmbH,
DE, part of Springer Nature 2024
M. Poppe, *Basic Electrodynamics in 6 Lessons*,
https://doi.org/10.1007/978-3-662-69143-4_1

1

"The path to the source is always against the stream." (*Arabian saying*)

Abstract Classical electrodynamics is still based on the same equations as in the late 19th century. In the meantime, however, their physical interpretation has changed substantially. Maxwell and his contemporaries did not doubt that all space was filled with a strange substance called ether. In his days, matter's electrical and atomic structure was utterly unknown, just like quantum mechanics and relativity. This chapter will summarise some of the key discoveries of the 20th century that changed the interpretation of Maxwell's equations. The resulting conjectures on electromagnetism will be stated at the end. They will form the basis for all the following chapters.

1.1 Charge and Fields

This book deals with a single property of matter called charge and its modification of nearby space. It also describes how this modification affects charged matter. The modification of the space surrounding charges is described by the terms electric field and magnetic field. The mathematical model of these fields, their relations among each other, and their relation to charges is called electrodynamics. The practical use of electrodynamics is electrical engineering.

Generally, a field is a function of some abstract space. In classical Physics, this space has three dimensions, each of which may be measured in meters. In this book, the term "field" shall be restricted to *measurable* functions of these three dimensions. This restriction has the following reason: All natural sciences (including electrodynamics) are empirical. Scientific progress is made by repeatedly comparing predictions of models with measurements. Models predicting properties of fields that cannot be measured can neither be verified nor falsified. Thus, such models do not contribute to scientific progress. Usually, they disappear after some time: The "ether" may be the most prominent example; the so-called "relativistic mass" (which misinterprets the effect of time dilation as a property of mass) is another one. The restriction of using only measurable fields follows Einstein's advice in the foreword: the number of quantities used in a theory should be as small as possible. This strategy was particularly successful. It led to the general theory of relativity.

1.2 **Some Results from Fundamental Research**

Hardly any electrical device works the way imagined by the early pioneers of engineering. Whereas most formulas used in the late nineteen hundreds are still valid, their interpretation has changed in the 20th century. Some of the scientific results leading to these changes will be outlined below:

E and B belong together

The following thought experiment illuminates the fact that different observers can perceive *apparently different* fields in the same place at the same time.

◾ Figure 1.1 shows the path of electrons in a vacuum chamber (here, a glass flask), visible as a pink glow due to collisions with the residual atoms in the vacuum. An external magnetic field causes the circular path of the electrons. It is, therefore, created by the Lorentz force $F = Qv \times B$. Now imagine that a spaceship with extragalactic scientists passes through the laboratory at exactly the speed at which the electrons move in the glass flask. If the scientists move at point P in ◾ Fig. 1.1 exactly parallel to the electrons at point Q, they will observe no Lorentz force because the velocity of the electrons appears to be zero for them. In the absence of alternative explanations, the aliens must conclude that the curvature of the electron orbit (for them: the acceleration of electrons at rest away from them) is caused by the Coulomb force $F = QE$. They, therefore, see the electron movement as proof for the presence of an electric field. The consequence of this thought experiment is as follows:

◾ **Fig. 1.1** Thought experiment on the presence and absence of magnetic fields. For the terrestrial observer, the electrons in the vacuum glass pass through a magnetic field on a circular path. The crew of a flying saucer flying at the speed of the electrons will find at point P that at point Q, there is no Lorentz force. (Foto: Marcin Bialek in commons.wikimedia.org)

1

■ **Fig. 1.2** The theory of
relativity in the original from
1905. (Facsimile [1])

891

3. Zur Elektrodynamik bewegter Körper;
von A. Einstein.

Daß die Elektrodynamik Maxwells — wie dieselbe gegen-
wärtig aufgefaßt zu werden pflegt — in ihrer Anwendung auf
bewegte Körper zu Asymmetrien führt, welche den Phänomenen
nicht anzuhaften scheinen, ist bekannt. Man denke z. B. an
die elektrodynamische Wechselwirkung zwischen einem Mag-
neten und einem Leiter. Das beobachtbare Phänomen hängt
hier nur ab von der Relativbewegung von Leiter und Magnet,

Observers with different velocities see different proportions of *E* and *B* fields
at the same place.

At the beginning of the 20th century Einstein analysed this phenomenon
and published the result under the title "Zur Elektrodynamik bewegter Koer-
per", (■ Fig. 1.2) today known as "Special theory of relativity" [2]. At its
core, it is an analysis of the transformation properties of Maxwell's equations and the resulting consequences. Einstein's following insights are cru-
cially important for modern understanding of classical electrodynamics:

❯ Coulomb Force and Lorentz Force

1. A consistent calculation of the movement of charge carriers always
   requires the use of the sum of both Coulomb force and Lorentz force
   for all frames of reference.
2. The fields *E* and *B* are different aspects of the same phenomenon. Their
   relative strengths appear to vary with the speed of the observer.

These two statements are two sides of the same coin: the close relationship
of the fields justifies the need always to use both. And the necessity of using
both fields proves their close relationship. Bluntly speaking: "One person's
B field is somebody elses *E* field."

Matter consists of free or bound charged particles

A branch of quantum mechanics, solid-state physics, has developed the picture of the electrical structure of solid matter (see e.g. [3]) shown in
■ Fig. 1.3. The stationary atomic nuclei (+) are surrounded by electrons
whose energies can only vary within certain ranges called energy bands
(shaded gray in ■ Fig. 1.3).

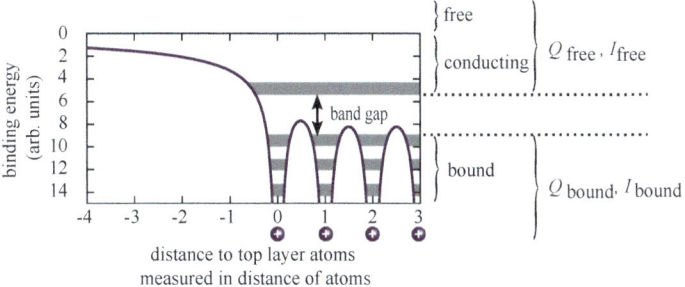

Fig. 1.3 Sketch of the spatial and energy structure of a solid according to the rules of quantum mechanics

In the early days of electrical engineering, free charges were assumed to be the only ones. Later, a growing number of observations indicated the presence of the solid bodies' electrical properties. But since no physical model could explain them, a distinction was made between "true" and "apparent" charges. Today, we know:

Electrical Properties of Solids

1. Solids consist of atomic nuclei and electrons only. The laws of electrodynamics and quantum mechanics apply to all of them.
2. The electrons can only assume certain energies organised in bands.
3. There are some energy bands in which the electrons can move almost freely within the solid.
4. Every electron is either free or bound. Consequently, the set of all electrons consists of two disjoint subsets.

In this book, no distinction will be made between truly free electrons and electrons that can move freely only within a certain material.

1

Field generation and field absorption are of the same nature

Today, classical electrodynamics is regarded to be a special case of quantum electrodynamics, i.e. its low energy density limit. Consequently, the laws of quantum electrodynamics also apply to classical electrodynamics. In quantum electrodynamics, charged bodies interact by an exchange of the quanta of the electromagnetic field (photons). Roughly speaking, the emission of photons describes the creation of a field, the effects of this field by their absorption.

Significant advances in theoretical physics arose from studies of symmetries. Symmetries deal with transformations that leave their object unchanged. For example, a rotation by 360° around any axis will leave a human body unchanged (maybe apart from being sick). Such a rotation is called symmetry operation or symmetry transformation. Generally, if a system remains unchanged under a certain transformation, this transformation is called a *symmetry transformation*. The system is then called *invariant for this symmetry transformation*.

Electrodynamic systems are believed to be invariant under the symmetry transformations shown in ◘ Fig. 1.4. These are

- *Charge conjugation, C,* i.e. swapping positive and negative charge: $C(Q) = -Q$.
- *Parity, P,* i.e. inverting the spatial coordinates. Examples are coordinates of points in space, $P(r) = -r$, forces $P(F) = -F$ and partial derivatives $P(\partial/\partial x) = -\partial/\partial x$. Axial vectors, i.e. those that consist of a vector product like $\tau = r \times F$ keep their sign under a parity transformation: $P(\tau) = +\tau$.
- *Time reversal, T,* i.e. the inversion of the time coordinate: $T(t) = -t$. As a result of this operation $T(\partial/\partial t) = -\partial/\partial t$, and thus $T(v) = T(\partial r/\partial t) = -v$, $T(a) = +a$ and $T(F) = +F$.

◘ **Fig. 1.4** Illustration of the symmetry operations charge conjugation *C, charge conjugation,* parity *P, parity* and time reversal *T, time reversal*

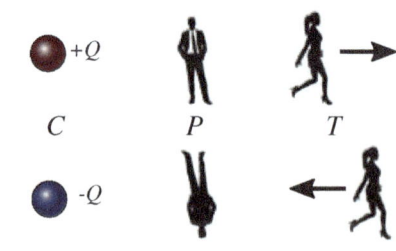

According to the Lueders-Pauli Theorem (CPT theorem, [4]) elementary processes occur in the same way if all three symmetry operations C, P and T are executed one after the other. It is of outstanding importance for electrodynamics because the so-called *CP invariance* has been investigated experimentally and confirmed with great precision. This invariance is equivalent to the following statement: "If one first inverts all spatial coordinates and then the signs of all charges, nothing changes." (see **Exploitation of CP Invariance in Particle Accelerators**). According to the CPT theorem, CP invariance ensures that electromagnetism is also invariant to a time reversal.

Coulomb force and Lorentz force are the classical limits of field absorption. For classical electrodynamics, this has the following consequences:

❯ The two Fields of Electrodynamics

1. The fields created by charges and those that exert on charges are identical.
2. Electrodynamics exclusively describes fields that manifest themselves through their effects on charges.
3. This implies that classical electrodynamics is about exactly two fields: E and B.

Maxwell's view of electromagnetism was quite different. He assumed the Coulomb force to result from a two-stage process. The charge would generate a "displacement of the ether" (D) and the ether would, in return, produce an electric field (E). Even many contemporary engineers assume electrodynamics to be a theory involving six fields rather than two. Today's view on the physical meaning of D, H, P and M will be worked out in the third lesson.

Exploitation of CP Invariance in Particle Accelerators

Many groundbreaking discoveries in the second half of the 20th century were achieved through experiments using particle accelerators. In the largest of them, *Large Electron Positron collider (LEP)* at CERN shown in ◼ Fig. 1.5, electrons in a vacuum tube were accelerated to speeds just below the speed of light and held on a circular path by strong magnets for up to an hour. In the same tube, positrons, the antiparticles of the electrons, circulated in the opposite direction. At the collision points, there were detectors which analysed the results of the matter-antimatter collisions.

1

⬛ **Fig. 1.5** Excerpt from the 27 km long *large electron collider, LEP* at CERN. One can see the deflection magnets (red), focusing magnets (blue), and with a closer look, the vacuum tubes for electrons and positrons (Photo: CERN)

Such experiments can only work because positrons in one direction follow precisely the same path as electrons in the other direction. Here, exactly means within 1 mm beam cross-section over a distance of a vast multiple of that accelerator circumference of 27 km. For a typical beam time of a quarter of an hour, this distance is twice the distance from the Earth to the Sun. The functioning proves the CP invariance of electromagnetism with extraordinary precision (approx. 1 mm / 300 million km): Charge conjugation (C) converts electrons into positrons, and parity (P) reverses the direction.

1.3 The Basic Assumptions of Electrodynamics

In light of the above research results, the following Statements inventory:

Conjecture 1.1 Electrodynamics is about a property of matter called *charge Q* and its interaction with the space surrounding it.

If the charge Q consists of so many individual charges that their distribution within a volume V can be viewed as continuous, it is helpful to use the *charge density ρ*[1]:

$$Q = \int_V \rho \, dV' \tag{1.1}$$

---

1   Integration variables are discussed in this book marked by an apostrophe (').

The product of charge density and speed of the charge carriers is called *current density* J:

$$J = \rho v \qquad (1.2)$$

Conjecture 1.2 Charges modify space. The modifications can be determined by the forces on the carrier of a charge Q_S acting as a probe: The Coulomb force $F = Q_S E$ indicates the presence of an electric field. The Lorentz force $F = Q_S v \times B$ acting on a carrier moving with the speed v indicates the presence of a magnetic field. Following a suggestion of Einstein's [2], the sum

$$F = Q_S(E + v \times B)$$

is called "electrodynamic force".

The electric field E is sometimes called the "field of the electric force". Since the magnetic field is also determined by its force, in this book, B will be called "field of the magnetic force".[2] Maxwell referred to B as *magnetic induction* [5]. This wording will be avoided because the term "induction" describes the effect of a changing magnetic field.

Conjecture 1.3 The creation and structure of the fields are determined by *Maxwell's equations* [5]. In SI units, they can be written in the following form:

1. $\nabla \cdot (\varepsilon_0 E) = \rho$ 2. $\nabla \cdot B = 0$
3. $\nabla \times E = -\frac{\partial}{\partial t} B$ 4. $\nabla \times (\mu_0^{-1} B) = J + \frac{\partial}{\partial t}(\varepsilon_0 E)$

The electric field constant ε_0 determines the magnitude of the electric field for a given charge density. The magnetic field constant μ_0 determines the magnitude of the field of the magnetic force.

---

2   The traditional term "magnetic flux density" will not be used. In all other natural sciences, a flux density is defined as the product of a density and a velocity.

1

The fact that one field constant is in the numerator and the other in the denominator is merely a convention. It has historical reasons only. A compact presentation the history and a very detailed list of original references to the historical development can be found in the script by Kirk T. McDonald [6].

The equations cited in the conjecture 1.3 are invariant for each of the three symmetry operations C, P and T, as can be seen by insertion (proof: solution of ◘ Problem 1.10). Consequently, Maxwell's equations in conjecture 1.3 obey the CPT theorem. A detailed discussion of the transformation properties can be found in [7].

1.4 **Problems**

1.1 What is the electrodynamic force, and who introduced it?

1.2 Can the effect of a magnetic field be re-interpreted into that of an electric field?

1.3 What is a conduction band?

1.4 In solid-state physics, the charge carriers are divided into four distinct groups. Which are they?

1.5 ◘ Figure 1.6 shows the quantum dynamical description of the fusion of an electron (e^-) with its antiparticle, the positron (e^+), to form a short-lived photon (γ), which decays into a pair of other elementary particles, so-called muons (μ^+, μ^-).

◘ **Fig. 1.6** Illustrating
◘ Problem 1.5: Creation
of a pair of elementary particles (*muons*). The fusion of an
electron (e^-) with a positron
(e^+) gives a very short-lived
photon (γ) which decays
immediately.

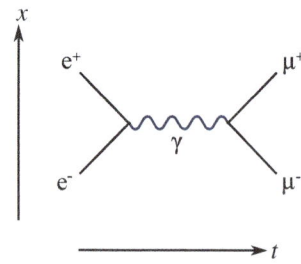

Please draw the changes in this picture that result from the operations C, P and T.

1.6 What is the immediate consequence of the CPT theorem given the CP symmetry of electrodynamics being proven by experiment? What is the consequence for the number of fields that electrodynamics deals with?

1.7 How can one compute the average speed for a set of carriers of charge from the average current density $< J >$ and which additional information is needed to do this?

1.8 Two of Maxwell's equations give the results of the divergences of the electrical field and the magnetic field. What do these formulas mean in visual terms?

1.9 Show that the *continuity equation*, $\partial\rho/\partial t + \nabla \cdot J = 0$ is a consequence of Maxwell's fourth equation (Ampere-Maxwell's law).

1.10 Please show that Maxwell's equations are invariant under the symmetry operations C, P and T.

1.5 **Solutions**

1.1 The electrodynamic force was defined by Einstein in 1905 as the sum of the Coulomb force and Lorentz force.

1.2 This is only possible for one point in space at a time. The Lorentz force disappears in a frame of reference, moving parallel to the magnetic field line touching this point. The force acting on a charge at this point must then be assigned to an electric field E.

1.3 An energy band is a region of binding energy in which electrons reside according to the laws of quantum mechanics. In a conduction band, a special case of an energy band, the binding energy is so weak that the electrons can move over distances of many atomic radii.

1.4 The groups are (1) electrons that have enough energy to leave the solid, (2) electrons trapped in the solid, but can freely move within the solid, (3) electron, which are bound to a single atom, and (4) atomic nuclei.

1

1.5 ◘ Figure 1.7 shows one particular solution. Other graphic ideas are possible and welcome.

1.6 Since the CP symmetry is fulfilled for electromagnetism, electromagnetic processes must also be invariant under time reversal. The fields that act on charges must be identical to those that generate charges. Consequently, charges can only produce the two fields: E and B.

1.7 The total number n of carriers of charge must be known. If each carrier has a charge q, the average current density can be written as

$$< J >= \frac{\int J dV'}{V} = \frac{\int \rho v dV'}{V} = \frac{q \sum_{i=1}^{n} v_i}{n} = q < v >$$

where $< v >$ is the average speed. So $< v >=< J > /q$ or, alternatively written as function of the total charge Q: $< v >=< J > n/Q$.

1.8 The equation $\nabla \cdot B = 0$ means the magnetic field has no sources. If Maxwell's equations are valid, there are no magnetic monopoles. It also implies that magnetic field lines have neither a beginning nor an end. They form closed loops.
  The equation $\nabla \cdot (\varepsilon_0 E) = \rho$ means that the electric field has sources and sinks if charges are present.

1.9 Starting with the *conjecture* 1.3.4 one finds

$$\nabla \times (\mu_0^{-1} B) = J + \frac{\partial}{\partial t}(\varepsilon_0 E) \qquad | \text{calculate deivergence}$$
$$\rightarrow \nabla \cdot (\nabla \times (\mu_0^{-1} B)) = 0 = \nabla \cdot J + \nabla \cdot \frac{\partial}{\partial t}(\varepsilon_0 E) \, | \text{swap}$$

◘ **Fig. 1.7** Illustrating Problem 1.5: The effect of symmetry operations C, P and T using the example of pair creation through electron-positron fusion

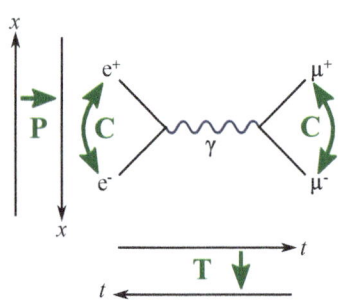

$$\to 0 = \nabla \cdot \boldsymbol{J} + \frac{\partial}{\partial t} \nabla \cdot (\varepsilon_0 \boldsymbol{E}) \mid \text{insert } conjecture \text{ 1.3.1}$$

$$\to 0 = \nabla \cdot \boldsymbol{J} + \frac{\partial \rho}{\partial t} \qquad \mid \text{done.} \tag{1.3}$$

The last formula is called the *continuity equation*.

Historically, the argument went precisely the reverse: the continuity equation was known. And it was trusted because it stated charge conservation at the infinitesimal level. Maxwell recognised that Ampere's law contradicted this equation and invented an additional term to fix it.

1.10 Charge conjugation C: $C(\rho) = -\rho$, and $C(\boldsymbol{J}) = -\boldsymbol{J}$ apply. Consequently, charge conjugation also flips the signs of $\boldsymbol{E}$ and $\boldsymbol{B}$. A look at the conjecture 1.3 shows that charge conjugation flips the signs on both sides of the equation, thus not affecting their validity.

Parity P: This transformation produces the following set of equations

1. $(-\nabla) \cdot [\varepsilon_0(-\boldsymbol{E})] = \rho$ 2. $(-\nabla) \cdot \boldsymbol{B} = 0$
3. $(-\nabla) \times (-\boldsymbol{E}) = -\frac{\partial}{\partial t}\boldsymbol{B}$ 4. $(-\nabla) \times (\mu_0^{-1}\boldsymbol{B}) = -\boldsymbol{J} + \frac{\partial}{\partial t}[\varepsilon_0(-\boldsymbol{E})]$,

which is identical to the original equations.

Time reversal T: Since the velocities change sign under time reversal, but the forces don't $T(\boldsymbol{J}) = -\boldsymbol{J}$ and $T(\boldsymbol{E}) = +\boldsymbol{E}$. However, $T(\boldsymbol{B}) = -\boldsymbol{B}$ because of $\boldsymbol{F} = Q\boldsymbol{v} \times \boldsymbol{B}$. Here, too, the original equations are reproduced after the transformation.

References

1. Faksimile siehe ▶ http://onlinelibrary.wiley.com/doi/10.1002/andp.19053221004/references
2. Einstein, Albert, Annalen der Physik Band 322, Nr. 10, Seiten 891ff, Leipzig (1905)
3. Konrad Kopitzki und Peter Herzog, Einfuehrung in die Festkoerperphysik, 7. Aufl., Springer Heidelberg 2017, ISBN 9783662535783
4. Gerhart Lueders, On the Equivalence of Invariance under Time Reversal and under Particle-Antiparticle Conjugation for Relativistic Field Theories, Kongelige Danske Videnskabernes Selskab, Matematisk-Fysiske Meddelelser 28 (5) 1-17 (1954)
5. James Clerk Maxwell, A Dynamical Theory of the Electromagnetic Field, Royal Society Transactions, Vol. CLV, 1865, p 459
6. Kirk T. McDonald, Forces on Magnetic Dipoles, Princeton University, New York 2018, ▶ http://kirkmcd.princeton.edu/examples/neutron.pdf
7. Juergen Schnakenberg, Elektrodynamik, Wiley-VCH Weinheim 2003, ISBN 978-3-527-40369-1

Charge and Current

Contents

© The Author(s), under exclusive license to Springer-Verlag GmbH,
DE, part of Springer Nature 2024
M. Poppe, *Basic Electrodynamics in 6 Lessons*,
https://doi.org/10.1007/978-3-662-69143-4_2

2

Abstract Electrical engineering is entirely based on the movement of charge carriers. Forces in electrical machines are forces on moving electrons. Accelerated charge carriers let antennas send and receive. And the diffusion of electrons into impurities in semiconductor crystals forms the basis of all transistor functions. In this chapter, the essential properties of the charge, i.e. its composition of elementary charges and its conservation, will be presented. The relationship between charge, current density, and current will also be laid out. It should also become clear which assumptions lead to Ohm's law and under which circumstances deviations occur.

2.1 Current and the Conservation of Charge

Charge conservation has been tested very accurately. It is most spectacularly observed in the pair creation of elementary particles. In a so-called bubble chamber charged particles leave a trail of small bubbles in a liquid, as shown in ◘ Fig. 2.1. If the chamber is exposed to a strong magnetic field, the curvature of the trails reveals the charge of the particles traversing the chamber. In decades of experimentation only pairs with a net charge of zero have been observed.

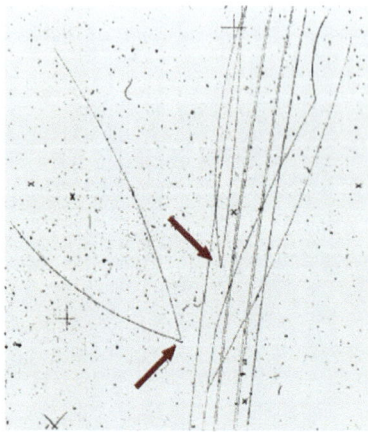

◘ **Fig. 2.1** Bubble chamber photo demonstrating charge conservation. Charged particles are always generated in pairs with a net charge of zero. This image [1] from 1964 shows the generation of pairs of charged particles at the points marked by arrows. The opposite curvatures in the magnetic field indicate that the particles have opposite charges

Current is based on motion, but it is not a vector

To start with, the connections between current density, current and an oriented area exposed to the current density shall be established (An oriented surface is a vector whose size is equal to the surface area and whose direction is perpendicular to the surface.).

The product of a charge density and the velocity of its carriers is called current density, J. It is, therefore, a vector pointing in the same direction as the carrier velocity. If this vector is multiplied by an oriented surface consisting of many individual elements da', the result turns out to be a change of charge with time: a loss of charge on one side of the surface and an equally large gain on the other side. This temporal change is called *current* and calculated in this

$$I = \frac{dQ}{dt} = \int_a J \cdot da' \tag{2.1}$$

way. The dot product in (2.1) selects the component of the current density, which is perpendicular to the surface and parallel to the surface vector. If both vectors are parallel; they indicate on which side the area the charge increases, as shown in ◘ Fig. 2.2. Note that the vectors of elements of a closed surface always point outwards.

The continuity equation: What is lost inside is added outside

◘ Equation (2.1) can now be used to determine a property of space that is valid at every point. This property is charge conservation at the infinitesimal level. For its mathematical description, the particular case of a stationary surface enclosing a volume V is considered. The following formula

◘ **Fig. 2.2** On the connection between current I and current density **J**: Where the area vector **a** and the current density are in parallel, the charge Q increases with time; where they are anti-parallel, the charge diminishes

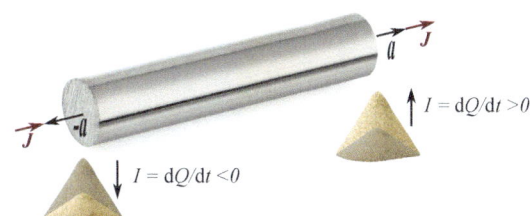

2

$$I_{\text{outwards}} = \frac{\partial Q_{\text{outside}}}{\partial t} = \oint_{\partial V} \boldsymbol{J} \cdot d\boldsymbol{a}' ,$$

describes the situation.[1] Here, the integration region ∂V stands for the area surrounding the volume V.

If no charge is lost, its increase outside of the volume V must be equal to the loss within the volume. Consequently,

$$\frac{\partial Q_{\text{outside}}}{\partial t} = -\frac{\partial Q_{\text{inside}}}{\partial t} = \oint_{\partial V} \boldsymbol{J} \cdot d\boldsymbol{a}'$$

should be fulfilled.

According to ◘ Eq. (1.1) $Q_{\text{inside}}/V = \rho$ is the average charge density. After inserting this into the above equation, in the limit $V \to 0$, one gets an equation which is valid for all points in space,

$$\frac{\partial \rho}{\partial t} + \nabla \cdot \boldsymbol{J} = 0 , \tag{2.2}$$

the *continuity equation*. It is the microscopic version of charge conservation. The contradiction between the continuity equation and Amere's law (remember ◘ Exercise 1.9) was the starting point of Maxwell's correction of this law. Today, conservation of charge is generally considered to be so self-evident that in formulations like "The current flows from a to b", it is unconsciously assumed (see ◘ Fig. 2.2).

A derivation that is complementary to the one just shown uses conservation of charge to *define* the term current density. For this purpose, a volume with arbitrarily moving boundaries is postulated to fulfil $0 = dQ/dt$ and cleverly transformed (see also the ◘ Problem 2.6 and its solution):

First, the charge is expressed as an integral over its density. Then, the derivative with respect to time is taken. The second integral of the result

$$\frac{dQ}{dt} = \frac{d}{dt} \int_V \rho \, dV' = \int_V \frac{\partial \rho}{\partial t} dV' + \oint_{\partial V} \rho_{\text{boundary}} \frac{dV'}{dt}$$

---

1   Here, the partial derivative is sufficient since the surface is at rest, and as a result, all derivatives contributing to I except $\partial Q/\partial t$ are zero.

describes that part of the change of charge, which is caused by the temporal change in the size of the volume, i.e. by dV'/dt. Here ρ_{boundary} is the charge density at the boundary.

With the help of Green's theorem, the integral over the moving boundary can be rewritten as

$$\oint_{\partial V} \rho_{\text{boundary}} \frac{dV'}{dt} = \int_V \nabla \cdot \left(\rho \frac{dr'}{dx} \right) dV' = \int_V \nabla \cdot (\rho v') \, dV'$$

i.e. as an integral over the entire volume. Here, from a mathematical point of view, v' is the velocity vector of an element of the volume. Physically, it may be interpreted as the velocity of the charge carriers.

The requirement to preserve the charge then reads

$$0 = \frac{dQ}{dt} = \frac{d}{dt} \int_V \rho dV' = \int_V \left(\frac{\partial \rho}{\partial t} \rho + \nabla \cdot (\rho v') \right) dV'.$$

As this equation is to hold for an arbitrary volume V, the integrand itself must be zero. With $J = \rho v$, the continuity ◻ Eq. (2.2) is always fulfilled.

❯ Charge Conservation and Current Density

Conservation of charge and current density are inextricably linked. Conservation of charge cannot be formulated at the infinitesimal scale without current density J. And the concept of current density only makes sense because the charge is a conserved quantity.

Kirchhoff's junction rule is a consequence of charge conservation

In the design of electrical systems, charge conservation appears in the form of the *junction rule*. This rule states: If no charge can be stored in an electrical node consisting of conductors with cross sections a_i, $(dQ_{\text{node}}/dt = 0)$, then

$$0 = \int_a J \cdot da' = \sum_i J_i \cdot a_i = \sum_i I_i \quad \text{(junction rule, original)}.$$

must be fulfilled.

If, however, the charge can be stored in a node, the rule does not apply. Consider, for example, a set of lines on a circuit board having a total parasitic

2

capacitance C_P against a fixed potential. Because of $dQ/dt = C dU/dt$, the junction rule has to be modified in the following manner:

$$\sum_i I_i + C_P \frac{dU}{dt} = 0 \quad \text{(junction rule with parasites)}.$$

The higher the frequency, the more important parasitic capacities.

2.2 Ohm's Law

In conductors, lattice oscillations slow down the charge transport by electrons

Ohm's Law [2] is a special case of Drude's law [3] which states that the current density in a body is proportional to the electric field, i.e

$$\boldsymbol{J} = \sigma \boldsymbol{E} \tag{2.3}$$

must be fulfilled. The proportionality factor σ is called *specific conductivity*.

◻ Equation (2.3) may be applied to the conductor shown in ◻ Fig. 2.2. It has has a cross-sectional area a and a length l. If it is exposed to tension² $U = E \cdot l$, then, with the help of $I = \boldsymbol{J} \cdot \boldsymbol{a} = Ja$, the well-known

$$U = \frac{l}{A\sigma} I = RI \tag{2.4}$$

appears. The factor R is called *resistance*. It is defined as the ratio of voltage to current. In most cases, it is only very weakly dependent on the voltage.

The proportionality of voltage and current gives a hint about the mechanism by which the solid slows down the charge carriers. To keep the argument simple, imagine the charge carriers as quasi-free-flying balls in the solid. Along the direction of the electric field, the formulas

$$QE = m_e \frac{dv}{dt} \qquad\qquad\qquad | \text{ 1. integration}$$
$$QEt = m_e v \,(+c_1) = m_e \frac{dx}{dt} \,(+c_1) \,|\, \text{2. integration with } c_1 = 0$$
$$\tfrac{1}{2} QEt^2 = m_e x \,(+c_2) \qquad\qquad | \text{ with } c_2 = 0$$

---

2   More details about voltages and potentials in the next chapter.

apply in this case. Here, $c_1 = 0$, means that after every impact, the new initial speed is zero and $c_2 = 0$ indicates that the the origin of the coordinate system is placed at the collision point.

First, stationary obstacles (ions of solid) are considered. Let x be the distance between two collisions. The final velocity v will then be reached after a certain time t. As the atoms do not move, one can assume that x does not depend on the field strength. Then, the time available to accelerate the charge carrier decreases like $t \sim 1/\sqrt{E}$. This decrease turns out to be a less-than-linear increase of the final velocity: If t is eliminated from the last two equations of the above system, the average velocity of the charge carriers and the current density emerges as

$$< v > = \frac{v_{end}}{2} = \sqrt{\frac{xqE}{2m_e}}$$

$$\rightarrow J = \rho \sqrt{\frac{xqE}{2m_e}} \qquad \text{(stationary obstacles)}.$$

One may conclude that if the movement of the solid atoms was negligible, $I \sim \sqrt{U}$ should be observed.

Alternatively, one can assume that the movement of the ions of the solid is entirely responsible for the slowing down of the electrons.[3] The time t between two collisions is then defined by the oscillation frequency $t = 1/(2f)$ of the ions (two collisions per oscillation). With $v_{end} = QE/(2m_e f)$ then follows

$$J = \rho \left(\frac{Q}{4fm_e} \right) E \quad \text{(fast moving obstacles)}, \tag{2.5}$$

i.e. the most commonly observed proportionality of current and voltage.

❯ Dynamic Origin of Ohm's Law

The validity of Ohm's law suggests that the movement of the ions is the only cause for the velocity reduction of the electrons.

The increase in resistance with temperature observed in many materials is now at least qualitatively understandable: As the temperature increases, the oscillation frequency must increase. Einstein suggests a quantum mechanical

---

3   You can imagine a jogger in a "horror tunnel" whose walls oscillate with large amplitude. So the jogger is repeatedly knocked over before he can pick up speed again.

2

approach [4] which supports this argument. According to this approach, the vibration frequencies of the ions in every spatial direction are multiples of the so-called *Einstein frequency* f_E. The vibration energies W_n are then

$$W_n = \left(n + \frac{1}{2}\right) hf_E, \quad \text{with} \quad n = 0, 1, 2, \ldots \tag{2.6}$$

with h being the *Planck constant*. Like photons, the oscillation frequencies are proportional to the oscillation energies. Now, increasing temperature is nothing other than increasing energy density. Consequently ◻ Eqs. (2.5) and (2.6) imply that the current density is inversely proportional to the absolute temperature T:

$$|\boldsymbol{J}| \sim \frac{1}{T} \leftrightarrow R \sim T,$$

Therefore, at room temperature, the resistance increases by about 1% per three degrees Celsius: $\Delta R/(R\Delta T) = 3.3\,10^{-4}/\,°C$. ◻ Table 2.1 shows that measured values are not the same. They are, however, of a similar order of magnitude for many metals. Significantly different values can occur with alloys. The temperature coefficients of semiconductors are negative: The increase in the number of charge carriers with temperature overcompensates their decreasing mobility.

◻ **Table 2.1** Temperature coefficient of the specific resistance at 20°C

Material	Ag	Au	Al	Cu	Fe	Hg
$\Delta R/(R\Delta T)$ in $10^{-4}/\,°C$	3,8	3,7	4,0	3,9	6,6	0,9

Requirements for the Conductivity of the Material for a Faraday Cage

◻ Figure 2.3 sketches a Faraday cage in a thunderstorm. Its protection can be equally beneficial for drivers and aircraft passengers. The key question is: "Which electrical properties are necessary for the material of the cage?" These properties shall be determined for a lightning current of $I = 1$ kA with the requirement of less than 10 Volts being generated inside the cage?

■ **Fig. 2.3** Image montages of the Faraday cage and the calculation the field strength variables used

To simplify the argument, we assume a spherical cage with a radius $r = 1$ m. It comprises a conductive layer with thickness δ and specific conductance σ. We also assume that the lightning has a finite diameter of $b_{\text{lightning}} = 0.1$ m and penetrates the entire cage. We set the lightning's direction as the z-axis.

With this information, the current density and, consequently, the field strength in the middle between the lightning's entry area and its exit area (at the "equator", if you imagine the entry area as the north pole and the exit area as the south pole) can determined as follows:

$$I = J_{\text{equ.}} \cdot a_{\text{equ.}} = 2\pi r J_{\text{equ.}} \delta \rightarrow E_{\text{equ.}} = \frac{I}{2\pi r \delta \sigma}.$$

Here, Drude's law (2.3) has been used to determine the strength of the field from the current density. Next, the variation of the current density and the field strength with the position on the cage are to be calculated. Let θ be the angle between a surface segment and the z-axis, then $E(\theta) = E_{\text{equ.}}/\sin\theta$ applies. Consequently,

$$E(\theta) = \frac{I}{\sin\theta \, 2\pi r \delta \sigma}$$

is the field strength along the path of the lightning's current. The voltage relative to the area of entry may now be obtained by integration:

$$U = 2 \int_{\theta_{min}}^{\pi/2} E(\theta')r d\theta' = \frac{I}{\pi\delta\sigma} \ln\left(\frac{\tan(\pi/4)}{\tan(\theta_{min}/2)}\right)$$

with $\theta_{min} = b_{lightning}/2r = 0.05$. The term $1/(\sigma\delta)$ is known as *sheet resistance* R_S. Because a square of a conducting layer always has the same resistance, regardless of size. For the parameters chosen here, the logarithm of the trigonometric functions gives a value of 3.45. Then,

$$U = \frac{R_S I}{\pi} \cdot 3.45 \approx R_S I$$

is a good estimate of the total voltage between lightning entry and exit. According to the specifications, the value of the sheet resistance $R_S \approx 10$ mΩ should not be exceeded. This value can easily be undercut by metal layers with $\delta < 1$ μm. For most applications, one may thus conclude: The limiting factor of the thickness of the material for a Faraday cage is not the maximum voltage that occurs, but rather the thermal stability, i.e. the melting of the metal in the event of insufficient heat dissipation at the impact site. The latter limits the value for the layer thickness.

The validity of Ohm's law is limited by thermodynamics

At very high electron speeds, the current is no longer proportional to the voltage. The proportion of kinetic energy transferred from the electrons to the solid increases. And this energy makes the difference. In conductors, the values for the kinetic energy that can be attributed to the drift velocity of the electrons are many orders of magnitude lower than those for the lattice ions. In semiconductors, however, this may no longer be the case. ◘ Figure 2.4 shows that, as a consequence, deviations from Ohm's law can be observed. In silicon, there is a final value for the velocity of approximately $v_{end} \approx 10^5$ m/s. The value is surprisingly close to the average thermal velocity of an ideal electron gas at room temperature.

■ **Fig. 2.4** Drift velocity of electrons in silicon as a function of field strength. A constant resistance would correspond to the dotted line. In contrast, measurements show a final velocity of about 10^5 m/s

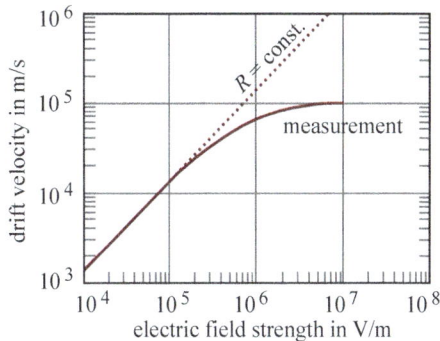

$$v_{e,rms} = \sqrt{< v_e^2 >} = \sqrt{\frac{3kT}{m_e}} = 1.16 \cdot 10^5 \text{ m/s},$$

which suggests the following conclusion:

❯ **The Thermal Limit of Om's Law**
Thermodynamics limits the validity of Ohm's law by limiting the average velocity of the electrons.

More detailed quantum mechanical calculations confirm this view (see e.g. [5,6]).

2.3 **Problems**

2.1 Let J be the current density at one point in a conductor. Find a simple expression for this of heat power dissipation density at this point. (You might start by analysing a very small volume of material, having a specific conductivity σ).

2.2 Which law is the basis of Ohm's law? Please give a formula as well.

2.3 Please name the hydrodynamic equivalent to the continuity equation of electrodynamics.

2.4 What is the general definition of a flux density or current density? And which of the following quantities obey this definition? (a) charge current density, (b) particle current density, (c) mass current density, (d) power density of a radiation source, (e) magnetic flux density?

2.5 If the law of conservation of charge is superior to Maxwell's equations, subordinate or equal?

2.6 ◘ Figure 2.5 sketches a a current passing a very small cube in the direction of x. Without using the continuity equation and without Gauss's theorem, show in the limit case $\Delta x \to 0$ the formula $\frac{\partial \rho}{\partial t} + \frac{\partial J_x}{\partial x} = 0$ applies.

2.7 It has been estimated that at room temperature, the resistance of a conductor increases by 1 % for every three degrees increase in temperature. How many degrees Celsius will it be at 100 °C for an increase of 1 %?

2.8 Resistance values can be determined using a setup usually referred to as *four-point method* and sketched in ◘ Fig. 2.6. The total current I is measured and also the voltage U_V between the contacts 1 and 2.

Please estimate the accuracy of the resistance determination, assuming that the contact resistances have values in the order of magnitude of $R_1 \approx R_2 \approx R_3 \approx R_4 \approx 10\ \Omega$, and that the internal resistance of the voltmeter has a value of $R_V \approx 1\ M\Omega$. The conduction zones immediately to the right and left of the resistance to be measured can be added to the contact resistors R_3 and R_4. Can you give a rule of thumb for the accuracy of the measurement method?

2.9 You know Ohm's law in the form $U = RI$. Are there circumstances under which $U = -RI$ must be used instead, and if so, which ones?

◘ **Fig. 2.5** Illustrating ◘ Problem 2.6: An infinitesimally small cube with current densities J_1 and J_2 coming in and out

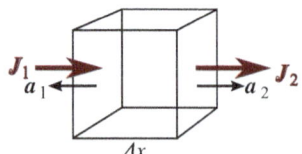

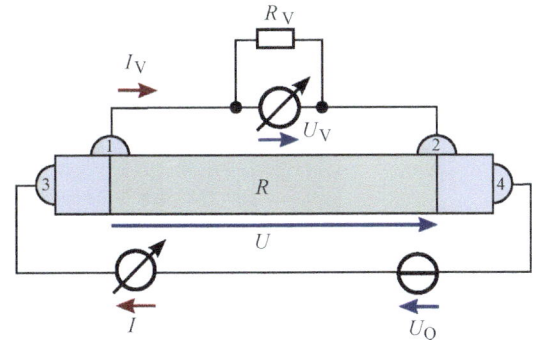

□ Fig. 2.6 Sketch of a four-point measurement to determine resistance here for the resistance R. The hemispheres indicate the four contact resistances. R_V is the internal resistance of the voltmeter

2.10 Carbon fibre aircraft are coated with a thin copper weave for protection against flashes of lightning. Please determine the sheet resistance of a copper grid consisting of square-shaped meshes. The weave consists of wires with radius r and distance d from wire centre to wire centre. Please investigate the case in which the lightning's current flows along one wire direction and compare the sheet resistance of this weave with the sheet resistance of a copper foil with the same mass per square meter.

2.4 Solutions

2.1 The cube-shaped volume $V = (\Delta x)^3$ shown in □ Fig. 2.7 shall be analyzed. It is assumed to be so small that the field $\boldsymbol{E}$ and the current density $\boldsymbol{J}$ are constant within the entire volume. The current penetrates vertically into one cube area. Now, one can conclude:

$$
\begin{aligned}
\boldsymbol{J} &= \sigma \boldsymbol{E} \\
I &= J \cdot \Delta x \cdot \Delta x \\
U &= E \cdot \Delta x = J \cdot \Delta x / \sigma \\
P &= U \cdot I = J^2 \cdot (\Delta x)^3 / \sigma \\
\rightarrow P/V &= J^2 / \sigma .
\end{aligned}
$$

Therefore, the heat power density is J^2/σ.

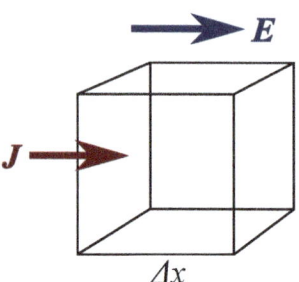

■ **Fig. 2.7** Illustrating Problem 2.1: Penetration of a charge density into a small volume $(\Delta x)^3$, exposed to an electric field $\mathbf{E}$

2.2 Ohm's law results from Drude's law, according to which the current density increases proportionally to the field strength (■ Eq. (2.3)).

2.3 The hydrodynamic law can be formulated by using mass density ρ of the water and the flux density $\mathbf{J} = \rho \cdot \mathbf{v}$, i.e. the product of mass density and flow velocity. In this case, the continuity equation looks like the one for the conservation of charge, but it states the conservation of mass.

2.4 A current density is always understood to be the product of a density and a speed. Examples are (a) charge density times charge carrier velocity, (b) particle density times particle speed, (c) mass density times speed, (d) energy density times the speed of light = power density, but not (e) the so-called magnetic flux density. The latter is based on a rather poor analogy between hydrodynamics and magnetostatics.

2.5 Historically, charge conservation was assumed first. But charge conservation contradicts Ampere's "law". Maxwell believed in charge conservation and extended Ampere's "law" to overcome the contradiction. One should rather speak of "Ampere's low-frequency approximation" than of "Ampere's law".

Logically, it's exactly the other way around: One can deduce the continuity equation from the Ampere-Maxwell law and Gauss's law for the electric field, but not vice versa. Consequently, Maxwell's equations are logically superior to the law of conservation of charge.

2.6 The net current into the volume, I_{in}, is the difference between the incoming current and the outgoing current. Linearisation leads to the result

$$I_{in} = -(a_1 \cdot J_1 + a_2 \cdot J_2) \quad | \text{ only components in } x - \text{direction}$$
$$= -a_{2x}(J_{2x} - J_{1x}) \quad | \text{ now linearise!}$$
$$= -a_{2x}\frac{\partial J_x}{\partial x}\Delta x \quad | \text{ divide by } V = a\Delta x$$
$$\frac{I_{in}}{V} = -\frac{\partial J_x}{\partial x} \quad | \text{ insert } Q = \rho V \text{ and current definition}$$
$$\frac{\partial \rho}{\partial t} = -\frac{\partial J_x}{\partial x} \quad | \text{ done,}$$

which is the one-dimensional version of the continuity equation.

2.7 Because the resistance increases in proportion to the absolute temperature, one can define a constant k with $R = kT$. Consequently, $\Delta R/\Delta T \approx dR/dT = k$. Then,

$$\Delta R/R \approx \Delta T/T.$$

Inserting a 1 % increase of themperature starting at 100 °C gives

$$\Delta T \approx 1\% \cdot 373 \text{ K} = 3.7 \text{ K},$$

or alternatively $\Delta T = 3.7 \,°C$.

2.8 The following voltages occur in the circuit:

$$U = (I - I_V)R \quad | \text{ tension at the object}$$
$$U = (R_1 + R_2 + R_V) \quad | \text{ the same tension, measured differently}$$
$$U_V = R_V I_V \quad | \text{ voltage atthe voltmeter}$$

Now, U and I_V can be calculated. The result

$$U_V = RI\left(1 - \frac{R_1 + R_2 + R}{R_1 + R_2 + R_V + R}\right) \approx RI\left(1 - \frac{R + 20\ \Omega}{R + 1\ M\Omega}\right)$$

shows, that for a value $R = 1\ k\Omega$, the accuracy is one per mille. As long as the internal resistance of the voltmeter is significantly greater than that of the contacts, the measurement errors are in the order R/R_V.

2.9 Reference directions can be freely assigned within a network. If, for a certain circuit element, these are antiparallel for the voltage and the current, the component equations must be $U = -RI$, $I = -CdU/dt$ and $U = -LdI/dt$. The reason is that whenever antiparallel reference directions are used, the following physical principle is violated: "When describing a system, all quan-

2

tities must be specified relative to the same coordinate system.".". Reference directions are nothing other than one-dimensional coordinate systems. Consequently, choosing them not to be parallel on a component is a physical modelling error that needs to be fixed by a minus sign in the component equation.

2.10 If the current flows along the direction of one set of wires, the wires placed transversely to this set do not contribute to the conductivity of the grid. Now, consider a square with side length l. A single wire has a $R_1 = l/(\sigma \pi r^2)$ resistance. In the square, there are $n = l/d$ parallel wires. This causes a reduction of the resistance by a factor of $1/n$. One obtains

$$R = \frac{R_1}{n} = \frac{l \cdot d}{\sigma r^2 \cdot l} = \frac{d}{\sigma \pi r^2} \,,$$

i.e. a result that is independent of the size of the square, as indicated by the fact that l has disappeared.

A square of copper foil with thickness δ and side length l has, also independent of this length, the resistance $R_S = 1/(\sigma \delta)$. Now δ is chosen so that the volumes (and therefore also the masses) of a foil and the mesh are the same, so

$$V_{\text{square}} = l^2 \delta = V_{\text{wires}} = n \cdot V_{\text{one wire}} = \frac{l}{d} \cdot \pi r^2 l \rightarrow \delta = \frac{\pi r^2}{d}$$

which results in the same sheet resistance

$$R_S = \frac{1}{\sigma \delta} = \frac{d}{\sigma \pi r^2}$$

for both the foil and one layer of the weave.

There is, however, a factor of two in favour of the foil's conductivity because the grid has two layers, one of which does not contribute to the conductivity perpendicular to its orientation. This factor of two is reduced to $\sqrt{2}$ in case of a current flow along the diagonal of the grid.

References

1. V. E. Barnes et al., Phys. Rev. Lett. 12, 204 (1964)
2. Georg Simon Ohm, Die galvanische Kette, T.H. Riemann, Berlin 1827
3. Paul Drude, Lehrbuch der Optik, Leipzig (1906)
4. Philipp Hofmann, Einfuehrung in die Festkoerperphysik, Wiley VCH, Stuttgart 2013, ISBN: 978-3-527-41226-6
5. Rainer Waser (Ed.), Nanoelectronics and Information Technology, 3. Ed., Wiley-VCH, New York 2012, ISBN: 978-3-527-40927-3
6. Harald Ibach und Hans Lueth, Festkoerperphysik, Springer Heidelberg 2009, ISBN 978-3-540-85794-5

Maxwell's Equations

Contents

© The Author(s), under exclusive license to Springer-Verlag GmbH,
DE, part of Springer Nature 2024
M. Poppe, *Basic Electrodynamics in 6 Lessons*,
https://doi.org/10.1007/978-3-662-69143-4_3

Abstract Classical electromagnetic field theory is completely described by Maxwell's equations. Consequently, one can rely on everything deduced from these equations. In this lesson, you will learn how the four Maxwell equations determine the formation and structure of electric and magnetic fields. You will see the field quantities D, P, H and M emerging from an application of Maxwell's equations to free and bound charges. And you will be able to trace the well-known macroscopic laws of electrical engineering back to Maxwell's equations.

3

3.1 Maxwell's Equations in Differential Form

Today, Maxwell's equations are not only considered the basis of Electrodynamics, but are an integral part of the *electrosweak interaction*. In this context, they form an indispensable building block for all attempts to design unified field theories. Accordingly, their validity can be trusted without restriction.

The laws summarised in Maxwell's equations of electromagnetic fields describe their causes and structure. The causes are charges and currents. They are (see Conjecture 1.3):

$$1.\ \nabla \cdot (\varepsilon_0 E) = \rho \quad\quad 2.\ \nabla \cdot B = 0$$
$$3.\ \nabla \times E = -\frac{\partial}{\partial t} B \quad\quad 4.\ \nabla \times (\mu_0^{-1} B) = J + \frac{\partial}{\partial t}(\varepsilon_0 E) \tag{3.1}$$

❶ Misleading Terminology

These equations are also referred to as *vaccum equations* suggesting non-validity in the presence of matter or *microscopic equations* suggesting validity at small distances only. None of these restrictions exist, as will be shown in this chapter.

The top two equations determine the *source structure* of the fields, as sketched in ◘ Fig. 3.1. The electric field has sources. Electrical charges are the starting points and the endpoints of electrical field lines. At the tiny scale, this feature is represented by the charge density ρ. The magnetic field has no sources. As a result, the number of magnetic field lines entering an arbitrarily shaped volume equals the number of field lines leaving this volume.

The two bottom equations describe the proportion of fields that closed lines can represent. It is also called the *vortex* or *rotational component* of the fields. Electric vortex fields go with the temporal change of magnetic fields.

◘ **Fig. 3.1** Illustration of
Maxwell's equations [1], pre-
sented in the same order as in
Eq. (3.1)

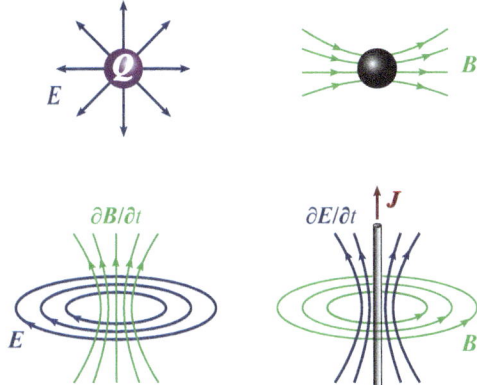

Magnetic fields are always rotational. Temporal changes in electric fields may cause them as well as currents. At the tiny scale, the currents are represented by current densities J.

The electric fields created by charges have sources and sinks. These fields can be superimposed by electric vortex fields caused by temporal changes in magnetic fields. The magnetic fields themselves are always pure vortex or rotational fields. The existence of electric vortex fields has a practical consequence:

❶ Limitations for the Mesh Rule

Kirchhoff's mesh rule [2] only applies in the absence of electromagnetic irradiation, because the electric field is only conservative in the static case.

The modified mesh rule, which takes into account radiation, is described in the solution of ◘ Problem 3.6.

3.2 Fields in Matter

Matter exposed to fields changes these fields as it creates electric or magnetic field contributions under the influence of "external" fields. These effects are usually described by the statement "matter can be magnetised or electrically polarised". These processes are therefore called *polarisation* or *magnetisation*

3

◘ **Fig. 3.2** Field balance in an electrically polarisable substance. An external electric field pulls the molecules apart. They become dipoles, weakening the field created from the outside.

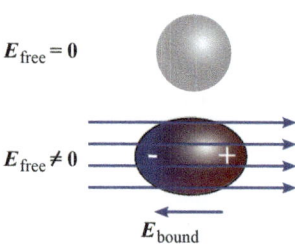

$$E_{\text{free}} = 0$$

$$E_{\text{free}} \neq 0$$

$$E_{\text{bound}}$$

As ◘ Fig. 3.2 shows, molecules with ionic or polar bonds are aligned to the external field if not influenced by other forces. This process creates many small dipoles whose field weakens the original field. This is called *polarisation.* Due to this process, the electric field inside a solid is composed of two contributions resp. fractions: the field contribution E_{free}, created by the charges outside the solid and the polarization field contribution of the solid E_{bound}:

$$E = E_{\text{free}} + E_{\text{bound}} \; . \tag{3.2}$$

This equation applies at every point in space.

When using ◘ Eq. (3.2), one should keep in mind that in most cases, the field contribution, which is due to polarised atoms only exists in the presence of an external field. This is one aspect of the following rule:

❯ Summation and Scale

Electric or magnetic field contributions are *indistinguishable fractions* of the complete fields, which must not be calculated separately and then added.

Another aspect is not as prominent:

Even isotropic materials change the directions of fields

When the materials are distributed in space in an inhomogeneous manner, polarisation may change the electric field's direction. Consequently, entire fields from different sources cannot simply be added. ◘ Equation (3.2) must be used *only at the microscopic level.*

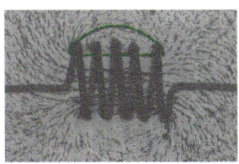 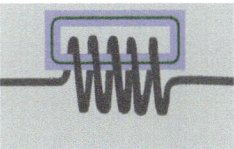

■ **Fig. 3.3** Counterexample to the addition of complete fields: The iron filings in the left picture show the course of the field generated by the coil. Introducing an iron core changes both the strength and the course and thus the length of the field lines

Forming Fields by Placing Ferromagnetic Materials

Magnetic fields also change their shape according to the distribution of magnetisable materials. ■ Figure 3.3 shows an example. The coil's magnetic field creates another magnetic field in the (previously unmagnetised) iron. At this moment, both mutate from fields to field fractions (field contributions). The sum of the two is the overall field B. In the terminology of solid-state physics, this means that the magnetic field fraction generated by free charges B_{free} aligns the bound currents of the iron so that they generate an additional field contribution B_{bound}. The result for the field of the coil-iron combination is not simply the sum of two macroscopic "fields" because iron is an almost isotropic material. It can amplify fields but not rotate them. Nevertheless, the iron, together with the coil, has created a field which has a different shape than that of the coil alone.

The electric field and the magnetic field are the sums of two contributions, each

Maxwell's equations are linear partial differential equations. Charge densities, current densities, and fields occur only to the first power. Consequently, the *principle of linear superposition* holds. ■ Figure 3.4 sketches the consequences: Given a set of charges and currents, one can either add them and calculate the fields from the sum, or one can calculate the fields for all charges and currents separately and then add the fields. The results will be the same. The principle of linear superposition works for fields

— always at the level of the differential equations,
— macroscopically in static fields in the absence of charge carriers,
— macroscopically with time-dependent fields in the absence of charge carriers, however with additional delay terms.

3

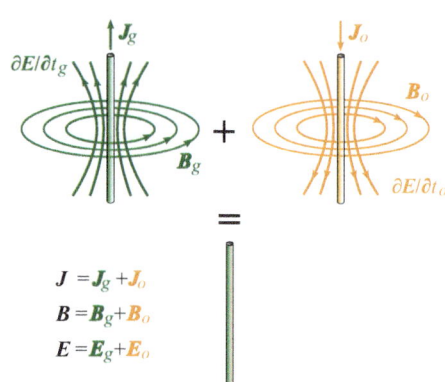

■ **Fig. 3.4** Illustration of the principle of linear superposition. Maxwell's equations apply to all non-overlapping sets of charges and currents separately and for any linear combinations. In the extreme case shown, both the currents and the fields are extinguished mutually

$$J = J_g + J_o$$
$$B = B_g + B_o$$
$$E = E_g + E_o$$

The solutions of the differential equations describe the macroscopic behavior of the fields and the charges. This includes the interaction of the charge carriers due to the fields and interactions between the fields mediated by charge carriers. Hence, at the macroscopic level, all fields and all charges are interconnected. The principle of linear superposition can only be applied in the example shown in ■ Fig. 3.4 at the macroscopic level if ideal current sources determine the currents and the conductors are mechanically stable.

Those differential equations which determine the behaviour of fields in the presence of matter may now be derived using the principle of linear superposition: At every point in space, the summations

$$\boldsymbol{B} = \boldsymbol{B}_{\text{free}} + \boldsymbol{B}_{\text{bound}} \quad \text{and} \quad \boldsymbol{E} = \boldsymbol{E}_{\text{free}} + \boldsymbol{E}_{\text{bound}} \tag{3.3}$$

apply. In addition, Maxwell's equations apply (3.1) for any set of charges, in particular

$$
\begin{array}{ll}
1a.\ \nabla \cdot (\varepsilon_0 \boldsymbol{E}_{\text{free}}) = \rho_{\text{free}} & 1b.\ \nabla \cdot (\varepsilon_0 \boldsymbol{E}_{\text{bound}}) = \rho_{\text{bound}} \\
4a.\ \nabla \times (\mu_0^{-1} \boldsymbol{B}_{\text{free}}) = \boldsymbol{J}_{\text{free}} + \frac{\partial}{\partial t}(\varepsilon_0 \boldsymbol{E}_{\text{free}}) & 4b.\ \nabla \times (\mu_0^{-1} \boldsymbol{B}_{\text{bound}}) = \boldsymbol{J}_{geb} + \frac{\partial}{\partial t}(\varepsilon_0 \boldsymbol{E}_{\text{bound}})
\end{array}
\tag{3.4}
$$

for bound (ρ_{bound}, $\boldsymbol{J}_{\text{bound}}$) and free ($\rho_{\text{free}}$, $\boldsymbol{J}_{\text{free}}$) charges and currents. The two systems of ■ Eqs. (3.3) and (3.4) can now be resolved in such a way that the entire fields are functions of the free charges and currents only. The result

$$
\begin{aligned}
\nabla \cdot \varepsilon_0 (\boldsymbol{E} - \boldsymbol{E}_{\text{bound}}) &= \rho_{\text{free}} \\
\nabla \times [\mu_0^{-1}(\boldsymbol{B} - \boldsymbol{B}_{\text{bound}})] &= \boldsymbol{J}_{\text{free}} + \frac{\partial}{\partial t}\varepsilon_0 \boldsymbol{E}_{\text{free}}
\end{aligned}
\tag{3.5}
$$

is usually known in the following form

$$\begin{aligned}
\nabla \cdot (\varepsilon_0 \mathbf{E} + \mathbf{P}) &= \nabla \cdot \mathbf{D} = \rho_{\text{free}} \\
\nabla \times (\mu_0^{-1} \mathbf{B} - \mathbf{M}) &= \nabla \times \mathbf{H} = \mathbf{J}_{\text{free}} + \frac{\partial}{\partial t} \mathbf{D}
\end{aligned} \tag{3.6}$$

and it is often referred to as *Maxwell's equations in matter* or *macroscopic equations*. In these equations $\mathbf{P}$ is called *polarisation* or (better) *dipole density*. The reason for this wording will be discussed more in the text explaining ◻ Fig. 4.13. The vector $\mathbf{D}$ stands for the so-called *displacement current* or simply *electrical displacement*. These expressions are reminiscent of the originally complete name *displacement of the ether*. In the nineteenth century, there was faith in the existence of an ubiquitous ether. The task of $\mathbf{D}$ was to shift or move it. The ether was assumed to react by the creation of the field $\mathbf{E}$.

$\mathbf{M}$ is traditionally called *magnetisation* or also *magnetic dipole density* (more on this in the text explaining ◻ Fig. 5.10). In the nineteenth century, $\mathbf{H}$ was called *magnetic field*, from the middle of the twentieth century onwards, *magnetic excitation* was also used.

A comparison of the ◻ Eqs. (3.5) and (3.6) shows that *Maxwell's equations in matter* are merely an application of Maxwell's ◻ Eq. (3.1) to the case of two disjoint sets of charges and currents. It doesn't even matter where exactly the dividing line is drawn. In conductors, free charge carriers are understood to be the electrons of the conduction band, in liquids, the ions dissolved in them, and in semiconductors, sometimes also temporarily occupied atomic orbitals. The comparison of (3.5) and (3.6) also reveals the physical nature of the quantities appearing the "macroscopic equations". These are summarized in ◻ Table 3.1.

◻ **Table 3.1** Meaning of vectors in the "macroscopic" equations [1]

Quantity	H	D	M	P
Meaning	$\mathbf{B}_{\text{free}}/\mu_0$	$\varepsilon_0 \mathbf{E}_{\text{free}}$	$\mathbf{B}_{\text{bound}}/\mu_0$	$-\varepsilon_0 \mathbf{E}_{\text{bound}}$
Caused by	Free currents	Free charges	Bound currents	Bound charges
Cause (symbol)	$\mathbf{J}_{\text{free}}$	ρ_{free}	$\mathbf{J}_{\text{bound}}$	ρ_{bound}

This table may be interpreted as follows: The value for $\mu_0 H$ at a certain point in space is the contribution to the field of the magnetic force that can be attributed to free currents. The value for $-P/\varepsilon_0$ is the contribution to the field of the electric force from bound charges, and so on.

The "microscopic" equations apply at the macroscopic scale

One may notice that this interpretation of the quantities appearing in Maxwell's equations in matter have little in common with Maxwell's view on electromagnetism. Another indication of the change of interpretation is the fact that the equations for bound charges, namely

$$\nabla \cdot \varepsilon_0 (E - E_{\text{free}}) \quad\quad = \rho_{\text{bound}}$$
$$\nabla \times [\mu_0^{-1}(B - B_{\text{free}})] = J_{\text{bound}} + \tfrac{\partial}{\partial t}\varepsilon_0 E_{\text{bound}} \;, \tag{3.7}$$

or

$$\nabla \cdot (\varepsilon_0 E - D) = -\nabla \cdot P \quad = \rho_{\text{bound}}$$
$$\nabla \times (\mu_0^{-1} B - H) = \nabla \times M = J_{\text{bound}} - \tfrac{\partial}{\partial t} P \tag{3.8}$$

were only published in the 2010s [3,4]. Their correctness is mandatory to keep both, the universal ◘ Eq. (3.1) and the "macroscopic equations" simultaneously valid (see also ◘ Problem 3.3 and its solution).

Why use "macroscopic" equations knowing that the "microscopic" ones are universally valid? Bluntly speaking, because the microscopic equations are useless as soon as the matter matters: The set of ◘ Eq. (3.1) relates all charges and all currents to the entire magnetic field and the entire electric field. The second half of this sentence is fine since engines use the entire magnetic field, and capacitors use the entire electric field. The first half of the sentence indicates the problem: Ampere meters measure free currents only. They do not measure the sum of free and bound currents. Similar statements apply to free and bound charges. Engineering needs equations that relate entire fields to fractional charges and currents. This is exactly what the macroscopic equations provide: They state the relations between free charges, free currents, and complete fields.

The term "auxiliary field" is misleading

Todays most often used term for the quantities H and D is "auxiliary field" (see e.g. [5]). Unfortunately, the quantities H and D are not *auxiliary*, and calling them *fields* leads to a semantical contradiction. An auxiliary variable is characterised by being part of an algorithm while not having any specific physical meaning. Quite in contrast, ◘ Table 3.1 shows that both of them do have a well-defined meaning. Also, it was shown that the "microscopic" equations are obeyed in the presence of matter. These equations define electromagnetism as a theory of exactly two fields, E and B. Consequently, H, D, M, and P cannot be physical fields. So what are they instead?

The use of the well-known formula $B = \mu_0(H + M)$ gives a hint: How can one add H and M knowing that one cannot add pears and apples? The only way to resolve this riddle is to realise that H and M are not fields of their kind, but they are both of the type B. Since the presence of the Lorentz force proves B to be the entire magnetic field, $\mu_0 H$ and $\mu_0 M$ must be fractions of this field.

The following example may illustrate the importance of distinguishing between an entirety and its contributions (its fractions). It is taken from hydrodynamics. In this example, indistinguishable parts are contributed from distinguishable sources, too.

Entirety and Contributions

The sensational headline shown in ◘ Fig. 3.5 is based on a fact that is undoubtedly true: At 301 m above sea level, near Pittsburg, the two the rivers Allegheny and Monongahela merge and form the Ohio River. One cubic meter of Ohio water gets half of its water molecules from the river Allegheny and the other half from the river Monongahela. In this sense, the statement that the Allegheny water within the Ohio River is only half as dense as it is in the Allegheny River is right: The rest of the volume is made up of Monongahela water. Whether a particular molecule comes from either river is fundamentally undetectable because individual molecules are indistinguishable. It is wrong to blame the decrease in density on the properties of water from the river Ohio. What is observed here is simply dilution. It is the merging of two indistinguishable parts (molecules from Allegheny and Monongahela) into a whole and its consequences.

A similar mistake is often made in magnetostatics. The magnetic (force-) field B is generated by two distinguishable currents, the subatomic ones ($\rightarrow M$) and those by the charge carriers with a larger radius of

3

Fig. 3.5 A sensational report, its truthfulness undermined by a confusion between totality and proportion. (Photo Wikiwand [6])

action ($\rightarrow$ **H**). At no location within the field, the cause can be determined experimentally. Because a Hall probe measures neither **M** nor **H**, but always **B**. Those who interpret the shapes of **H** and **M** as results of their individual properties make the same mistake as the journalist who attributes the halving of the density of water to the properties of the water from the river Allegheny and not as a result of the merger.

Just as the rivers Allegheny and Monongahela join to become the Ohio River, the magnetic field is the sum of two contributions from two sources, bound and free currents. Once added, the shares can no longer be separated. Even if the Atlantic exclusively consisted of Allegheny water, all ships would stay afloat. Water is water, no matter what source it is fed from, and a magnetic field is a magnetic field, regardless of whether free or bound currents generate it. Consequently, **E** and **B** are fields, all other vectors occurring in the "macroscopic equations" are fundamentally indistinguishable shares of them.

Relative field constants simplify formulae

The difference $E - E_{\text{bound}}$ may be re-interpreted as *modification of the electric field* E by the presence of matter. Since this is material-specific, it is best represented by a new variable ε_r, the *relative permittivity*. Its meaning results from the following reformulation of the field difference $E - E_{\text{bound}}$:

$$E_{\text{free}} = (E - E_{\text{bound}}) = \varepsilon_r E . \tag{3.9}$$

◼ Equation (3.9) *defines* the material-specific variable ε_r. ◼ Figure 3.6 shows two examples.

In most cases, ε_r is simply a constant, the *relative permittivity* or the *relative dielectric constant*. It indicates by which factor the substance *weakens* the electric field from free charges. This case is shown in ◼ Fig. 3.6 on the left. In solids, whose molecules or crystals cannot deform in the opposite direction to the outer field, ε_r is described by a transformation matrix because these materials change the direction of the electric field. Such materials are referred to as being *electrically anisotropic*.

The difference between the field B and the field contribution created by magnetisation, B_{bound} can also be interpreted as a modification of the magnetic field generated by the free currents. In this case,

$$B_{\text{free}} = B - B_{\text{bound}} = \mu_r^{-1} B \tag{3.10}$$

is the choice. The quantity μ_r is called *relative magnetic permeability*. For solids whose molecules cannot align (anti-)parallel to the outer magnetic field μ_r can be represented by a matrix because these materials change the direction of the magnetic field. Such materials are called *magnetically anisotropic*.

◼ **Fig. 3.6** Definition of ε_r: One can get a new vector E_{free} from an old one, E in two ways: Either by subtracting a vector E_{bound} or by transforming E. An elongation is shown on the left, a rotation by the angle α on the right

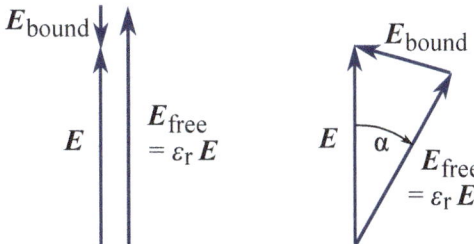

3

As long as field contributions from free charges and currents are present, μ_r and ε_r may be uses without any restrictions. Neither certain properties nor certain quantities are required for the material as long as μ_r and ε_r are not restricted to be constant numbers [1]. However, large quantities of substances imply that ε_r is almost independent of the position within the solid. And for isotropic materials, there is no need to use matrices.

The following simple rule sums it up:

Using ε_r and μ_r

The equations for fields in materials can be obtained from the "microscopic" equations by the substitutions

$$\begin{aligned}
\varepsilon_0 E &\rightarrow \varepsilon_0 \varepsilon_r E \\
\mu_0^{-1} B &\rightarrow (\mu_0 \mu_r)^{-1} B \\
\rho, J &\rightarrow \rho_{\text{free}}, J_{\text{free}}.
\end{aligned} \tag{3.11}$$

Here, ε_r and μ_r^{-1} generally substance-specific matrices. Care must be taken to place μ_r^{-1} directly in front of B and ε_r directly in front of E.

Often, the abbreviations $\varepsilon_0 \varepsilon_r = \varepsilon$ and $\mu_0 \mu_r = \mu$ are used. In the presence of matter (see Conjecture 1.3) the following set of equations

$$\begin{aligned}
&1.\ \nabla \cdot (\varepsilon E) = \rho_{\text{free}} \qquad 2.\ \nabla \cdot B = 0 \\
&3.\ \nabla \times E = -\frac{\partial}{\partial t} B \qquad 4.\ \nabla \times (\mu^{-1} B) = J_{\text{free}} + \frac{\partial}{\partial t}(\varepsilon E)
\end{aligned} \tag{3.12}$$

emerges. It looks astoundingly similar to the original Maxwell equations.

The equations for fields in matter do not apply at subatomic distances

The distinction between free and bound charges or currents only makes sense if volumes significantly larger than atomic orbitals are considered. Otherwise, quantum effects cannot be neglected. And within such orbitals, electrons are almost free. Consequently, one has to conclude: The use of E_{free}, B_{free}, D, H, P, M, μ_r and ε_r is unsuitable at atomic distances.

In contrast, E and B retain meaning.

3.3 The Integral Field Equations

While the differential equations provide the basic understanding of electro-dynamics, their integral form is much more useful for technical applications. Next, these integral equations shall be presented together with their relations to the Maxwell's equations. The following conventions shall be used: The symbol $\oint$ stands either for a closed line or for a closed area. The surface of an enclosed area A is denoted by ∂A, the surface of a volume V by ∂V.

Coulomb's law follows from Gauss' theorem for the electric field

Gauss' theorem connects a volume integral over the divergence of a vector field to the flow integral of this field through the volume's surface: The *Gauss theorem for the electric field* is

$$\oint_{\partial V} \varepsilon_0 \boldsymbol{E} \cdot \mathrm{d}\boldsymbol{a}' = \int_V \nabla \cdot (\varepsilon_0 \boldsymbol{E})\, \mathrm{d}V' = Q. \tag{3.13}$$

The left equal sign in (3.13) applies to any vector field, the one on the right only to the electric field. ◘ Figure 3.7 illustrates the meaning of this theorem. The electric field of a point-like charge is perpendicular to the spherical surface. It is, therefore, parallel the infinitesimal surface element vectors $\mathrm{d}\boldsymbol{a}'$. Because the surface of the sphere has the size $a = 4\pi r^2$, the strength of the electric field diminishes with the square of the distance:

$$\boldsymbol{E} = \frac{Q\boldsymbol{r}}{4\pi\varepsilon_0 |\boldsymbol{r}|^3} \tag{3.14}$$

Together with conjecture 1.2, Coulomb's law follows.

Seen from a distance, the sphere shown in ◘ Fig. 3.7 looks like the centre point for all electric field lines. Now consider the limiting case of tiny volumes, $V \to 0$ for the sphere. Then

◘ **Fig. 3.7** Illustration of Gauss' theorem. A charge at the centre of a sphere generates an electric field, which is perpendicular to the surface of the sphere

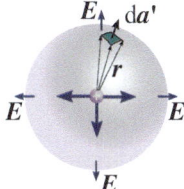

$$\lim_{V \to 0} \frac{Q}{V} = \lim_{V \to 0} \left(\frac{1}{V} \int_V \nabla \cdot (\varepsilon_0 \boldsymbol{E}) \, dV' \right) = \nabla \cdot (\varepsilon_0 \boldsymbol{E}) \tag{3.15}$$

emerges. Now, the charge density is $\rho = Q/V$, so that Gauss' theorem for the electric field describes the same phenomenon as Maxwell's first Eq. (3.1.1). In the presence of matter, ■ Eq. 3.11

$$\oint_{\partial V} \varepsilon \boldsymbol{E} \cdot d\boldsymbol{a}' = \int_V \nabla \cdot (\varepsilon \boldsymbol{E}) \, dV' = Q_{\text{free}} \,,$$

where in general, $\varepsilon = \varepsilon_0 \varepsilon_r$ must be placed within the integral.

Gauss' theorem for the magnetic field excludes magnetic monopoles

Gauss' theorem for the magnetic field is

$$\oint_{\partial V} \boldsymbol{B} \cdot d\boldsymbol{a}' = 0 \tag{3.16}$$

and can be related to the second Maxwell Eq. (3.1.2) with similar manipulations as before:

$$\oint_{\partial V} \boldsymbol{B} \cdot d\boldsymbol{a}' = 0 \Leftrightarrow \nabla \cdot \boldsymbol{B} = 0 \tag{3.17}$$

It states that the magnetic field has no sources. Consequently, there are no individual magnetic charges (monopoles). In field line images, this feature shows up in the fact that each volume has the same number of incoming and outgoing field lines. Also, magnetic field lines are closed.

The Faraday-Henry law combines two effects in a single formula

The law of magnetic induction, also called the Faraday-Henry law according to its discoverers, is

$$\oint_{\partial a} \boldsymbol{E} \cdot \mathrm{d}\boldsymbol{s}' = -\frac{\mathrm{d}\Phi_B}{\mathrm{d}t} = -\frac{\mathrm{d}}{\mathrm{d}t}\left(\int_a \boldsymbol{B}\mathrm{d}\boldsymbol{a}'\right) \tag{3.18}$$

and describes both the interaction between electric and magnetic fields, and the effect of magnetic fields on moving charges According to Richard P. Feynman, this is "the only case in physics where a single formula describes two completely different effects [7]." A mathematical analysis of the formula shall now check this statement:

According to the rules of vector analysis, taking the derivative of the integral in ◘ Eq. (3.18) for an arbitrary field $\boldsymbol{B}$ gives

$$\frac{\mathrm{d}}{\mathrm{d}t}\left(\int \boldsymbol{B}\mathrm{d}\boldsymbol{a}'\right) = \int \left\{\frac{\partial \boldsymbol{B}}{\partial t} + \boldsymbol{v}\left(\nabla \cdot \boldsymbol{B}\right) - \nabla \times (\boldsymbol{v} \times \boldsymbol{B})\right\} \cdot \mathrm{d}\boldsymbol{a}' .$$

From a mathematical point of view, $\boldsymbol{v}$ is the speed of the surface elements $\mathrm{d}\boldsymbol{a}'$. Physically, it may be interpreted as the movement of charge carriers.

Since the field of the magnetic force (the "magnetic flux density") has no source,

$$\frac{\mathrm{d}}{\mathrm{d}t}\left(\int \boldsymbol{B}\mathrm{d}\boldsymbol{a}'\right) = \int \left\{\frac{\partial \boldsymbol{B}}{\partial t} - \nabla \times (\boldsymbol{v} \times \boldsymbol{B})\right\} \cdot \mathrm{d}\boldsymbol{a}' \tag{3.19}$$

remains in this case. ◘ Equation (3.19) shows that the law of induction in its macroscopic form describes two different effects: the $\partial \boldsymbol{B}/\partial t$ term indicates that a change of magnetic field strength plays a role. This part is a necessary ingredient for the formation of electromagnetic waves. The time dependent term forms the theoretical foundation of all electric motors.

For a closer examination, the line integral in ◘ Eq. (3.18) is turned into a surface integral using Stoke's theorem. A comparison of the result

$$\oint_{\partial a} \boldsymbol{E} \cdot \mathrm{d}\boldsymbol{s}' = \int_a (\nabla \times \boldsymbol{E}) \cdot \mathrm{d}\boldsymbol{a}'$$

with ◘ Eq. (3.19) then yields

$$\nabla \times \boldsymbol{E} = -\frac{\partial \boldsymbol{B}}{\partial t} + \nabla \times (\boldsymbol{v} \times \boldsymbol{B}) . \tag{3.20}$$

3

▪ Equation (3.20) shows that the electric field generated by induction is a pure vortex field. The first two terms of this equation are identical to the third of Maxwell's Eq. (3.1.3). The two outer terms can be derived from the Lorentz force:

$$
\begin{aligned}
\boldsymbol{F} &= Q\boldsymbol{v} \times \boldsymbol{B} \\
\rightarrow \boldsymbol{E} &= \boldsymbol{v} \times \boldsymbol{B} \\
\rightarrow \nabla \times \boldsymbol{E} &= \nabla \times (\boldsymbol{v} \times \boldsymbol{B})
\end{aligned}
\tag{3.21}
$$

Note that ▪ Eq. (3.21) involves the division by a charge Q. If no charge is present, this means dividing by zero. Therefore, the following restriction applies.

❗ The Law of Induction Assumes the Presence of Free Charges

▪ Equation (3.18) may only be used in the presence of charge carriers. Otherwise, the formula

$$
\oint_{\partial a} \boldsymbol{E} \cdot \mathrm{d}\boldsymbol{s}' = -\int_{a} \frac{\partial \boldsymbol{B}}{\partial t} \cdot \mathrm{d}\boldsymbol{a}'
$$

which is equivalent to the third Maxwell equation must be used.

In technical applications like transformers or electrical machines, charge carriers are provided by conductors.

The Ampere-Maxwell law describes the generation of magnetic fields

The fourth law is the one of Ampere and Maxwell:

$$
\oint_{\partial a} (\mu_0^{-1}\boldsymbol{B}) \cdot \mathrm{d}\boldsymbol{s}' = I + \int_{a} \frac{\partial(\varepsilon_0 \boldsymbol{E})}{\partial t} \mathrm{d}\boldsymbol{a}'
\tag{3.22}
$$

The current I occurring in this law is the one which passes through the surface a. Again, the relationship to Maxwell's equations can be found with the help of Stokes' theorem,

$$
\oint_{\partial a} (\mu_0^{-1}\boldsymbol{B}) \cdot \mathrm{d}\boldsymbol{s}' = \int_{a} (\nabla \times (\mu_0^{-1}\boldsymbol{B})) \cdot \mathrm{d}\boldsymbol{a}' .
$$

By replacing

$$I = \int_a J \cdot \mathrm{d}a'$$

all terms in ▪ Eq. (3.22) can be turned into surface integrals:

$$\int_a (\nabla \times (\mu_0^{-1} B)) \cdot \mathrm{d}a' = \int_a \left\{ J + \frac{\partial(\varepsilon_0 E)}{\partial t} \right\} \cdot \mathrm{d}a' \tag{3.23}$$

Since this equation must apply to any surface a, the integrands must also be equal:

$$\nabla \times (\mu_0^{-1} B) = J + \frac{\partial(\varepsilon_0 E)}{\partial t} \tag{3.24}$$

This law is, therefore, equivalent to the fourth Maxwell Eq. (3.1.4).

In case $\partial(\varepsilon_0 E)/\partial t = 0$, ▪ Eq. (3.24) reduces to the so-called Ampere law

$$\oint_{\partial a} (\mu_0^{-1} B) \cdot \mathrm{d}s' = I , \tag{3.25}$$

which, strictly speaking, is only a partial law, as it is only valid in the absence changing electric fields. It may be regarded as the low-frequency approximation to the Ampere-Maxwell law.

3.4 Refraction of Field Lines

Both electric and magnetic fields change direction at material boundaries. This phenomenon is called *refraction of field lines*. For closer examination, it is helpful to decompose the field vectors into one component parallel to the material surface and one perpendicular to it.

As indicated in ▪ Fig. 3.8, the component of the electric field parallel to the surface remains unchanged as a consequence of the Faraday-Henry law. This can be seen by an investigation of the path connecting the points P_1 to P_4 in ▪ Fig. 3.8. Consider the case of an ever-shrinking distance between the points P_1 and P_4, as well as P_2 and P_3. while keeping the distance Δx between the points P_1 and P_2 finite. Then,

Fig. 3.8 Refraction of electric field lines at a material transition

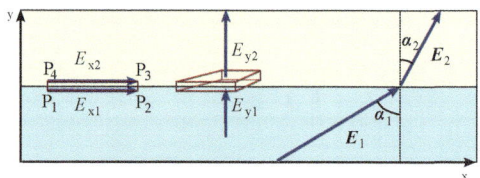

3

$$\oint_{\partial a} E \cdot ds' = -\frac{d}{dt}\left(\int_a B \cdot da'\right)$$

$$\to (E_{x1} - E_{x2})\,\Delta x = 0$$

$$\to E_{x1} = E_{x2}.$$

because the area integral must approach zero if the area tends to zero.

For an investigation of the vertical component, Gauss' law for the electric field, ■ Eq. (3.13) may be used. In the presence of matter,

$$\oint_{\partial V} \varepsilon E \cdot da' = \int_V \nabla \cdot (\varepsilon E)\,dV = Q_{\text{free}}.$$

is the choice. Now consider an infinitesimally thin volume as indicated in ■ Fig. 3.8. Only the top and bottom areas significantly contribute to the surface integral. Denoting their corresponding surface vectors by A and $-A$, one gets in the limit of an infinitesimally thin volume V

$$\left(\varepsilon_2 E_{y2} - \varepsilon_1 E_{y1}\right) A = 0$$

$$\to \varepsilon_1 E_{y1} = \varepsilon_2 E_{y2}$$

i.e. an expression that shows that the component perpendicular to the surface shrinks at the same rate as ε_r grows.

Finally, the expressions for the two field components can be combined in a single formula

$$\frac{\tan\alpha_1}{\tan\alpha_2} = \frac{E_{x1}/E_{y1}}{E_{x2}/E_{y2}} = \frac{\varepsilon_1}{\varepsilon_2}, \tag{3.26}$$

which is known as *law of refraction for electric field lines*.

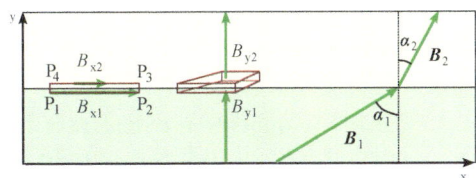

Fig. 3.9 Refraction of magnetic field lines at a material transition

The behaviour of magnetic fields at material boundaries can be determined (see ■ Problem 3.10) from the Ampere-Maxwell law and Gauss' law. ■ Figure 3.9 shows the result: a change in the surface-parallel field component and an unchanged vertical component. The combined result is

$$\frac{\tan \alpha_1}{\tan \alpha_2} = \frac{B_{x1}/B_{y1}}{B_{x2}/B_{y2}} = \frac{\mu_1}{\mu_2}, \tag{3.27}$$

the so-called *law of refraction for magnetic field lines.*

To sum it up: At the surface of a material, the component of a magnetic field parallel to the surface and the vertical component of an electric field change. The results of these complementary properties are the similarly looking formulas for the refraction laws of both fields. These laws of refraction apply to both static fields and slowly oscillating fields.[1]

3.5 Problems

3.1 Are Maxwell's "microscopic" ■ Eq. (3.1) valid in the presence of matter?

3.2 There are only four Maxwell equations, but in the presence of matter, six equations have to be obeyed. How many equations would have to be fulfilled if, within the set of bound charges and currents, one would distinguish between electrons and atomic nuclei?

3.3 Which of the following quantities have different meanings or properties depending on the presence or absence of matter: B, M, H, E, D, P?

---

1   The frequencies must be small enough for atomic relaxation times to be significantly shorter than the duration of one period.

3

3.4 Please restore Maxwell's "microscopic" equations from the systems of ◻ Eqs. (3.5) and (3.7).

3.5 If one wants to use the law of induction, one needs to calculate the line integral $\oint \boldsymbol{E} \cdot d\boldsymbol{l}$. Is it necessary to place the path elements $d\boldsymbol{l}$ parallel to the electric field lines?

3.6 Kirchhoff's mesh rule is $\sum_i U_i = 0$. Please suggest a modification of this rule for a circuit board that is exposed to electromagnetic radiation.

3.7 Assume an electric field to have the form

$$\boldsymbol{E} = E_0 r_0^2 \left(\frac{-y}{r^3}, \frac{x}{r^3}, 0 \right)$$

with $r^2 = x^2 + y^2 + z^2$ at the time $t = 0$. E_0 and r_0 are constants. Is this field created by charges or by magnetic fields?

3.8 Show that the electric field of a charged sphere has no vortex component.

3.9 A capacitor whose electrodes have circular areas with radius $r = 10$ cm and a mutual distance of $d = 50$ μm has an initial voltage of $U_0 = 5$ V. It is discharged via a resistor whose resistance has a value of $R = 100$ kΩ. The value for the relative permittivity of the dielectric is $\mu_r = 2$. Also $\varepsilon_r = 10$. Please calculate the field $\boldsymbol{B}$ between both capacitor plates as a function of the radius and the time. The radii may be smaller or larger than those of the capacitor electrodes. How does the magnetic field depend on ε_r? How does the electric field depend on time?

3.10 Please derive the law of refraction for the field lines of the field of the magnetic force $\boldsymbol{B}$ ("magnetic flux density").

3.11 In deriving the refraction laws for field lines of the electric field, it was assumed that the value for the surface integral $\int f \, da'$ vanishes if the area a tends to zero. Can you confirm this assumption or even explain it?

3.6 **Solutions**

3.1 Yes, the "microscopic" equations are valid in the presence of matter. But they relate the entirety of all charges and currents to the entirety of the fields. But only the free fraction of all charges and currents is accessible to measurement. Consequently, these equations are useless in the presence of matter.

3.2 For each new subset of charges and currents, two equations are added. The subdivision of the bound charges and currents each leads to a total of eight equations.

3.3 In a vacuum, the quantities H and D coincide with the electric and the magnetic field. In the presence of matter, they are only field fractions (contributions). They are no longer fields. Consequently, Maxwell's equations no longer make any structural specifications (poles, ...).

The quantities M and P have non-zero values in the presence of matter, only. In the absence of free currents or charges, they coincide (except for constants depending on the system of units) with the fields B and E. Otherwise they are field contributions with no structural specifications by Maxwell's equations.

The physical fields B and E always mean the same thing, and they must fulfil the structural requirements of Maxwell's equations under all conceivable circumstances.

3.4 One adds the equations for the electric field from different sources

$$
\begin{aligned}
\nabla \cdot \varepsilon_0 (E - E_{\text{free}}) &= \rho_{\text{bound}} \\
+ \nabla \cdot \varepsilon_0 (E - E_{\text{bound}}) &= \rho_{\text{free}} \\
\rightarrow \nabla \cdot \varepsilon_0 (2E - E_{\text{free}} - E_{\text{bound}}) &= \rho_{\text{free}} + \rho_{\text{bound}} \\
\rightarrow \nabla \cdot \varepsilon_0 E &= \rho
\end{aligned}
$$

and for the field of the magnetic force ("flux density"):

$$
\nabla \times [\mu_0^{-1}(2B - B_{\text{free}} - B_{\text{bound}})] = J_{\text{free}} + J_{\text{bound}} + \frac{\partial}{\partial t}\varepsilon_0 (E_{\text{bound}} + E_{\text{free}})
$$
$$
\rightarrow \nabla \times [\mu_0^{-1} B] = J + \frac{\partial}{\partial t}\varepsilon_0 E .
$$

3

QED. The two remaining Maxwell equations are independent of the presence of matter. They need no further investigation.

3.5 No, it is not necessary to calculate this integral along the field lines. Because the dot product between the path element $d\boldsymbol{l}$ and the field vector $\boldsymbol{E}$ has a value that is identical to the one obtained without vector calculation when choosing a path along the field lines: $\boldsymbol{E} \cdot d\boldsymbol{l}_{\text{arbitrary}} = Edl_{\text{field line}}$. It is not necessary to calculate the integral along a field line, but in most practical cases, it will be easier. If the shape of the field lines is unknown, the calculation is almost always impossible without numerical aids.

3.6 Kirchhoff's mesh rule follows from the absence of a vortex component in the electrostatic field. According to the law of induction (Faraday-Henry law), time-dependent fields can have a vortex contribution. Let Φ_B denote the magnetic flux through the area spanned by the mesh, then

$$\sum_i U_i + \frac{d\Phi_B}{dt} = 0 \quad \text{(mesh rule including irradiation)} \tag{3.28}$$

is the extension one was looking for.

3.7 The divergence of the field is calculated as a test for the presence of sources. The result

$$\nabla \cdot \boldsymbol{E} = E_0 r_0^2 \left(3\frac{yx}{r^5} - 3\frac{xy}{r^5} + 0 \right) = 0$$

shows that the field is a pure vortex field. The temporal change of a magnetic field, therefore, causes it.

3.8 The electric field of a sphere was calculated to be $\boldsymbol{E} = \boldsymbol{r}Q/(4\pi\varepsilon_0 |\boldsymbol{r}|^3)$. The rotation of this field is

$$\nabla \times \boldsymbol{E} = E_0 r_0^2 \left(-3\frac{yz}{r^5} + 3\frac{zy}{r^5}, -3\frac{zx}{r^5} + 3\frac{xz}{r^5}, -3\frac{xy}{r^5} + 3\frac{yx}{r^5}, \right) = (0, 0, 0),$$

as it was to be shown (The calculation may be shortened by using polar coordinates.).

3.9 Except for edge effects, the electric field is constant between the plates and practically absent outside. The magnetic field can be calculated using the Ampere-Maxwell law (3.22). If we call α the distance to the centre of the disk, then, a concentric magnetic field is between the plates, which must fulfil

$$\oint_{\partial a} (\mu_0^{-1}\mu_r^{-1} \boldsymbol{B}) \cdot \mathrm{ds}' = I + \int_a \frac{\partial(\varepsilon_0 \varepsilon_r \boldsymbol{E})}{\partial t} \cdot \mathrm{da}' \to 2\pi\alpha B = \mu_0 \mu_r \varepsilon_0 \varepsilon_r \frac{\partial E}{\partial t} \pi\alpha^2$$

Taking the derivative of the latter equation and inserting the time dependence of the electric field gives a magnetic field

$$B = -\frac{U_0 \mu_0 \mu_r \alpha}{R \, 2\pi r^2} \mathrm{e}^{-t/\tau} \quad , \quad (\alpha < r)$$

which becomes stronger towards the edge. Surprisingly, it does not depend on the electrical properties (ε_r) of the dielectric. The ε is cancelled out by the $1/\tau$ term the inner derivative of $exp(-t/\tau)$.

For radii larger than r, the line integral becomes independent of μ_r, because the line now runs outside the dielectric. The magnetic field then decreases according to

$$B = -\mu_0 \frac{U_0}{R} \frac{1}{2\pi\alpha} \mathrm{e}^{-t/\tau} \quad (\alpha \geq r)$$

In other words, outside the capacitor, the magnetic field is precisely as large as the field around the wires connecting the capacitor. The initial current $I_0 = U_0/R$, produces a field $B = \mu_0 I/(2\pi\alpha)$ of the same strength.

The value for the capacitance is now $C = \epsilon_0 \epsilon_r (\pi r^2)/d = 55.6$ nF. The time constant τ has a value $RC = 5.56$ ms. This gives the electric field the strength whose value may be calculated as

$$E = \frac{U_0}{d} \cdot \mathrm{e}^{-t/RC} = 10^5 \frac{\mathrm{V}}{\mathrm{m}} \cdot \mathrm{e}^{-t/(5.56 \text{ ms})}$$

3.10 Gau's theorem applied to the volume shown in ▣ Fig. 3.9 volume gives

$$\oint_{\partial V} \boldsymbol{B} \cdot \mathrm{da}' \approx (B_{y2} - B_{y1})a = 0 \to B_{y1} = B_{y2},$$

for the vertical component. Here, a is the area size of the top of the cuboid.

3

The Ampere-Maxwell law, ⬛ Eq. (3.22), in the presence of matter, ⬛ Eq. (3.11) applied to the path shown on the left of ⬛ Fig. 3.9 gives

$$\oint_{\partial A} (\mu^{-1} \mathbf{B}) \cdot d\mathbf{s}' = \int_A \mathbf{J} \cdot d\mathbf{a}' + \int_A \frac{\partial(\varepsilon \mathbf{E})}{\partial t} \cdot d\mathbf{a}' \quad | \text{ consider} |P_1 - P_4| \to 0$$

$$\to \left(\frac{B_{1x}}{\mu_1} + \frac{B_{2x}}{\mu_2} \right) |P_1 - P_2| = 0 \qquad\qquad | \text{ divide } / |P_1 - P_2|$$

$$B_{1x}\mu_2 = B_{2x}\mu_1$$

because the enclosed surface approaches zero. Consequently, the surface integrals in the Ampere-Maxwell law disappear.

The rest is geometry:

$$\frac{\tan \alpha_1}{\tan \alpha_2} = \frac{B_{x1}/B_{y1}}{B_{x2}/B_{y2}} = \frac{\mu_1}{\mu_2} .$$

QED.

3.11 This can be justified with the mean value of the function f over the surface:

$$<f> = \int_a f \, da' / a \qquad\qquad | \text{ definition of a mean value}$$

$$\to \int_a f \, da' = <f> \int_a da' = <f> a \quad | \lim_{a \to 0}$$

$$\to \int_a f \, da' = 0 \qquad\qquad | \text{ as long as } <f> \text{ is finite}$$

In other words, the integral over an area is written as the product of the area and the average value for the function over this area. The integral must disappear with the surface as long as this average is finite.

References

1. Martin Poppe, Pruefungstrainer Elektrotechnik, 4. Auflage, Springer Heidelberg 2022, ISBN 978-3-662-65001-1
2. Gustav Kirchhoff, Vorlesungen ueber mathematische Physik, 3. Band, Teubner Leipzig 1891, ► https://archive.org/details/vorlesungenberm01plangoog/page/n7/mode/2up
3. Andrzej Herczynski, Am. J. Phys. 81, 202 (2013); ► https://doi.org/10.1119/1.4773441
4. Gonano Zich and Mussetta, Progress In Electromagnetics Research B, Vol. 64,

5. David J. Griffiths, Electrodynamics, 5. Edition, Cambridge University Press 2023, ISBN 9781009397759
6. New Flame sinking off Europa Piont, ▶ www.wikiwand.com
7. Richard P. Feynman, The Feynman Lectures on Physics: The Definitive and Extended Edition, Vol.2, Addison Wesley 2005, ISBN 0-8053-9045-6

First Special Case: Static Electric Fields

Contents

© The Author(s), under exclusive license to Springer-Verlag GmbH,
DE, part of Springer Nature 2024
M. Poppe, *Basic Electrodynamics in 6 Lessons*,
https://doi.org/10.1007/978-3-662-69143-4_4

Abstract Central terms in electrodynamics can be traced back to the properties of static electric fields. After reading this chapter, you will understand the connections between the electric field, potential, voltage and energy. You will also know how the use of these terms emerges from the properties of the static electric field. You will recognise the interplay between field energy, potential energy, and the intrinsic electrical energy of matter. You will be familiar with some of the techniques used to calculate electric fields be it by solving the equations of Poisson and Laplace or be it the use of Green's function. Also included in this lesson are mirror charges and the approximation method of multipole expansion. You will recognise that the methods of classical electrical engineering become invalid at subatomic distances.

4

4.1 Energy and Potential

Getting familiar with the term *potential* is worth the effort:
- The difference between two potentials leads to the concept of voltage, which permeates all electrical engineering.
- The potential is a scalar quantity whose derivative gives all three components of the electrical field vector. Therefore, its determination is often the simplest option to determine the electric field.
- The potential is a measure of potential energy and is a connecting quantity between electrodynamics and mechanics.

In the static case, the electric field has no vortex component because $\nabla \times \boldsymbol{E} = -\partial \boldsymbol{B}/\partial t$ reduces to $\nabla \times \boldsymbol{E} = 0$. This static electric field is a pure source field. Therefore, it is easier to determine than time-dependent fields.

Two forms of energy must be distinguished: the *potential energy of a charge carrier in the electric field* and the *energy content of the field itself*. The law of conservation of energy inextricably links both: The energy content of a field grows to the extent that work has to be done to produce it by shifting charges against each other.

Charges give potential energy to each other

To move a charge carrier from a starting point to a target point, work must be done. The energy gain[1] is given by

---

[1]   The minus sign in ◼ Eq. (4.1) means that the energy is gained when moving against a force.

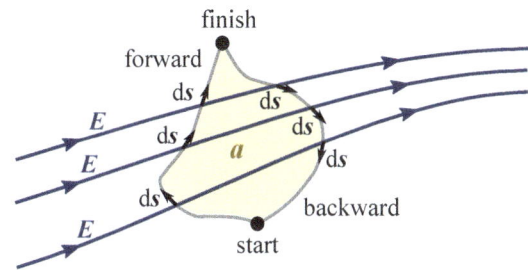

■ **Fig. 4.1** Closed curve in the static electric field E, which is interpreted as the sum of two different paths between points a and b. According to Stokes' theorem, the energy gain of a charge carrier is $-\oint QE \cdot ds = 0$ along the overall curve if the the electric field is a pure source field on the surface a

$$W_{pot} = \int\limits_{start}^{finish} -F \cdot ds' = -Q \int\limits_{start}^{finish} E \cdot ds' . \tag{4.1}$$

Since the vortex component of the static electric field is zero, the energy gain depends, except for the charge Q, on the field and from the start and endpoints only. The latter can be proven with the help of Stokes' theorem. Consider a closed path as shown in ■ Fig. 4.1 that leads through the starting point and the endpoint. The gain in potential energy obtained by moving through the entire path gives the energy difference between the "forward path." and "backward path". For the curve ∂a enclosing the area a Stokes' theorem reads

$$Q \oint\limits_{\partial a} E \cdot ds' = Q \int\limits_a (\nabla \times E) \cdot da' = 0,$$

which means that the absence of a vortex component ($\nabla \times E = 0$) one can assign a potential to every point in space. Therefore, W_{pot} is called *positional energy* or also *potential energy* of the charge carrier.

The work that is necessary to move a point charge Q_1 from a considerable distance up to a distance r_{12} to another charge, Q_2, results from ■ Eqs. (4.1) and (3.14) to be

$$W_{pot} = \frac{Q_1 Q_2}{4\pi\varepsilon_0 r_{12}} . \tag{4.2}$$

If a third charge is added, it must be moved against both the field of the first charge and against that of the second one. The generalisation to n point charges is then

$$W_{\text{pot}} = \frac{1}{4\pi\varepsilon_0} \sum_{i=1}^{n} \left(\sum_{k=i+1}^{n} \frac{Q_i Q_k}{r_{ik}} \right) \tag{4.3}$$

where the index range of the second summing ensures that each combination of indices only occurs once and that there is no interaction between a charge carrier and itself.

Figure 4.2 shows two examples for charge configurations, both consisting of a single charge Q and two further charges, each half in size, and held at a fixed distance A. The variation of the total potential energy is shown as a function of the distance x between the large charge and the centre of gravity of the two small ones according to ◘ Eq. (4.3). If $x = 0$, the large charge lies in the middle between the two small ones. In both cases, the total potential energy does not tend to zero for large x. It approaches the potential energy that the two small charges give each other. This energy is called *internal energy* of the system of two charges.

If all charges are lined up, the potential energy W_1 has poles at the x values corresponding to the distance between the large charge and one of the small ones being zero. If the connecting line between the charge carriers is perpendicular to the distance to the large charge, there is no pole because all charges remain separated, as shown in ◘ Fig. 4.2. The difference between the total

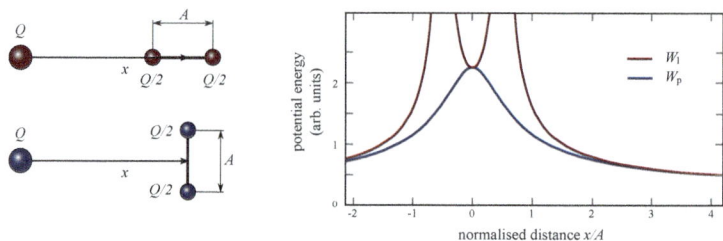

◘ **Fig. 4.2**   Two configurations, each consisting of three charge carriers (left) and their total energy (right). The charge carriers are viewed as point-like. Two of the charge carriers each have a fixed distance A. In the upper case, they are all lined up and give the total energy W_1 (l like line), standing in the bottom the connecting lines are perpendicular to each other and result in W_p (p like perpendicular)

potential energy and the internal energy is also called *interaction energy*. The above example may be generalised as follows:

❗ Internal Energies and Interaction

A calculation of the interaction energy can only be performed without knowledge of the internal structure of the interacting partners if the distance between them is significantly larger than their size. In that case, the partners may be approximated by pointlike charges.

Potential: a scalar quantity determines the electric field vector

The potential energy of a charge carrier at a point b, divided by its charge is called *electric potential* $\Phi(b)$:

$$\Phi(b) = \int_{\text{somewhere}}^{b} E \cdot ds' \tag{4.4}$$

The term "somewhere" as a lower limit indicates that the definition of potential involves a certain freedom: The definition of Φ does not specify the starting point of the charge carrier's path. Most of the time, but not always, one assumes that the starting point was far away from all other charge carriers.

The potential in the vicinity of an individual charge carrier is a function of the location, which determines how much energy a second charge carrier needs to approach the first charge carrier up to this location.

The potential has a similar meaning for the static electric field as the height (more precisely, the product of the height h and the acceleration due to gravity: $\Phi_{\text{gravity}} = gh$) in the earth's gravitational field. No matter which of the routes shown in ◘ Fig. 4.3 is chosen, the energy gain is always the same:

$$\Delta W = mg\Delta h = m\Delta\Phi_{\text{gravity}}.$$

The difference in the potential at two different points a and b is called *voltage* or *tension* between these points:

$$U_{ab} = -\int_{a}^{b} E \cdot ds' = (\Phi(a) - \Phi(b)). \tag{4.5}$$

4

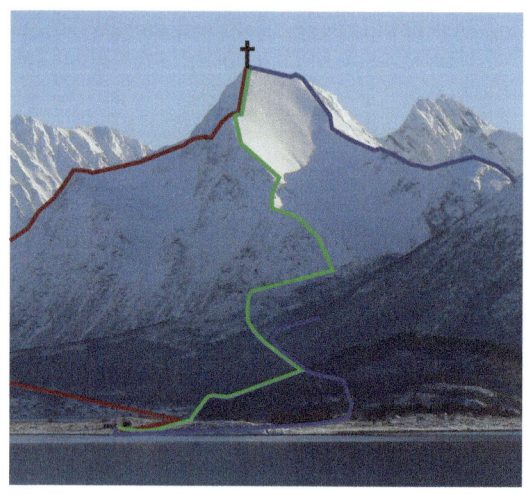

■ **Fig. 4.3** Various hiking routes to a mountain peak in northern Norway. The gain in potential energy is the same on all routes because the gravitational field has no rotational component

Given a specific charge, the voltage defines what energy is required or released to move its carrier from the first point to the second. In contrast to the potential, the voltage between two points is defined without freedom of choice.

Using a potential Φ significantly simplifies the calculation of electric fields. This becomes apparent if a complementary definition of the potential is used. The electric (vector) field E can can be written as a gradient of the scalar function Φ,

$$E = -\nabla\Phi \tag{4.6}$$

which suffices to determine all three components of the electric field. The equivalence of the ■ Eqs. (4.4) and (4.6) can be shown as follows: The energy gain dW along a piece of the path $d\mathbf{r}$ is always determined by the force F to be overcome. This means for a carrier of the charge Q that

$$dW = -Q\mathbf{E} \cdot d\mathbf{r} = -Q(E_x dx + E_y dy + E_z dz)$$
$$d(W/Q) = -\mathbf{E} \cdot d\mathbf{r} = -(E_x dx + E_y dy + E_z dz)$$

must be valid.

Let $\Phi = W/Q$ be the *potential* and assume that this function is differentiable and unique for every point in space, its total differential

$$\mathrm{d}\Phi = \left(\frac{\partial\Phi}{\partial x}\mathrm{d}x + \frac{\partial\Phi}{\partial y}\mathrm{d}y + \frac{\partial\Phi}{\partial z}\mathrm{d}z\right) = \nabla\Phi \cdot \mathrm{d}\boldsymbol{r}$$

looks very similar to that of gain in energy. A comparison of the last two equations then gives ◘ Eq. (4.6). Consequently, a static electric field can be written as the gradient of a potential function. In the dynamic case, when it acquires a vortex fraction, ◘ Eq. (4.6) is no longer applicable for $\nabla \times (\nabla f) = 0$.

◘ Equation (4.6) justifies the freedom of choice of the starting point in the definition of the electric potential (4.4): A change in the starting point increases or decreases the potential by a constant. This constant disappears under differentiation. So, the starting point can be "somewhere".

The field energy can partly be transferred to dielectrics

The energy contained in an electric field can be derived from the work needed to charge a capacitor (see f. ex. [1]). In this manner, the energy density turns out to be

$$w = \frac{W}{V} = \frac{\varepsilon_0\varepsilon_\mathrm{r}}{2}\boldsymbol{E}^2\,, \tag{4.7}$$

where ε_r is the relative permittivity of the dielectric.

For the capacitor shown in ◘ Fig. 4.4, for example, the polarizability of the dielectric triples the energy required for charging. ◘ Equation (4.7) leads to an answer to the question of how the energy is distributed: Since ε_r is just the factor by which the dielectric decreases the strength field and $\boldsymbol{E}$ is the field present in the material ("external field minus polarisation"),

◘ **Fig. 4.4** A film capacitor. Less than one-third of the energy stored in it is in the electric field $\boldsymbol{E}$, more than two-thirds are stored in the polarised molecules of the dielectric. (Image: WIMA GmbH & Co. KG)

$$w_{\text{capacitor}} = \frac{\varepsilon_0 \varepsilon_r}{2} \boldsymbol{E}^2$$

$$w_{\text{field}} = \frac{\varepsilon_0}{2} \boldsymbol{E}^2$$

$$\rightarrow w_{\text{field}}/w_{\text{capacitor}} = 1/\varepsilon_r,$$

is the result.

❯ Energy in Dielectrics

A dielectric introduced into an electric field takes up a significant part of the field energy so that the proportion of energy stored in the field itself amounts to $1/\varepsilon_r$, only.

The energy stored in the dielectric is exactly the energy needed for its polarisation - as required by conservation of energy.

The field energy sets limits to classical electrodynamics

In the context of the special theory of relativity, Einstein demonstrated that mass is a form of energy. The following consideration shows that, as a consequence, the laws of classical electrodynamics cannot be applied to very small objects: The energy of a point-shaped source diverges. Therefore, according to the laws of classical electrodynamics, its mass must also grow beyond all limits. This will be shown next.

The energy stored in the electric field outside a charged sphere of radius R is

$$W = \int\limits_{\text{outside}} w \, dV' = \int\limits_{R}^{\infty} w \, 4\pi r^2 dr',$$

where w is the energy density of the electric field. Using ◘ Eq. (3.14) it can be determined to be

$$w = \frac{1}{2}\varepsilon_0 \boldsymbol{E}^2 = \frac{Q^2}{32\pi^2 \varepsilon_0 r^4}.$$

The resulting total energy

$$W = \frac{Q^2}{4\pi\varepsilon_0 R},$$

diverges for $R \rightarrow 0$. Because of the equivalence of mass and energy ($W = mc^2$) the above formulas can no longer be used for very tiny objects. For the radius R_e of an electron, the limit is

$$R_e = \frac{Q^2}{4\pi\varepsilon_0 m_e c^2} \approx 2.8fm$$

because then, the electric field alone is as heavy as its creator. This minimum radius corresponds approximately to that of a light atomic nucleus. It is many orders of magnitude above the value experiments with electrons suggest. Consequently, the limit provides no information about the actual size of an electron. It proves the necessity of a modification of classic electrodynamics when analysing nuclear processes.

4.2 Field Calculation Using Gauss and Coulomb Integrals

Gauss' theorem helps for simple geometries

The use of Gauss' theorem for the calculation of electric fields is simple if it is known from symmetry that the field vectors E is parallel to the surface elements da. Then, the dot product $E \cdot da$ reduces to the product of the amounts Eda.

Doping Requirements for a Solar Cell

Gauss' theorem for the electric field can be used to determine the field strengths within a solar cell. Solar cells are semiconductor components in which differently contaminated areas border each other. A zone almost free of mobile charge carriers forms near the junction, the so-called depletion zone. The zone contains charged, immobile ions but no mobile electrons. Within this zone, electricity can be generated in the following manner. A photon releases an electron from its bond. Then, the electron leaves the depletion zone by following the electric field. The wider the depletion zone, the greater the probability of light-to-electricity conversion becomes. Therefore, it is crucial to know the relationship between the technology parameters and the width of the depletion zone. The depletion zone should be as wide as possible for a given voltage.

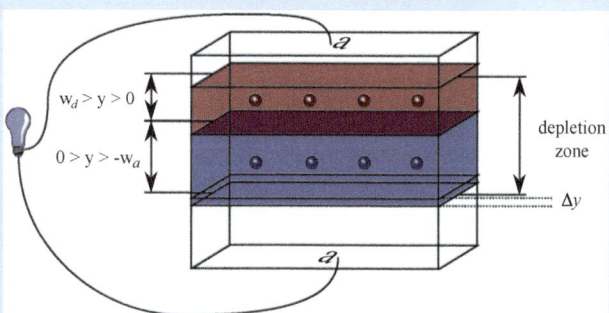

⬛ **Fig. 4.5** Immobile charge carriers (ions) in the depletion zone of a semiconductor PN junction. Gauss' theorem is applied to volume elements of size $a \Delta y$

⬛ Figure 4.5 shows a geometry of the depletion zone. It contains negatively charged ions on the lower side ($y < 0$) and positively charged on the high side ($y > 0$). Beyond the depletion zone, the material is conductive and therefore practically field-free. If Gauss' theorem is applied to the cuboid of thickness Δy and frontal area a as sketched in ⬛ Fig. 4.5, then

$$\oint \boldsymbol{E} \cdot \mathrm{d}\boldsymbol{a}' = E_y a \quad \text{and} \quad Q = \rho V = \rho a \Delta y,$$

where the charge density ρ is equal to the density of impurity atoms multiplied by the charge of an electron, $\rho = -e\,n_a$. The cuboid always has its lower edge at the lower end of the depletion zone, at $y_{\mathrm{low}} = -w_a$. For a cuboid of any thickness, $\Delta y = y + w_a$. For negative values of y, Gau's theorem (3.13) may be written as

$$\varepsilon E_y a = \rho a(y + w_a) \rightarrow E_y = -\frac{e n_a}{\varepsilon}(y + w_a),$$

leading to a field $\boldsymbol{E} = (0, E_y, 0)$ which disappears at the bottom of the depletion zone. It reaches its greatest negative value at $y = 0$, i.e. at the junction. For positive y values

$$E_y = \frac{e n_a}{\varepsilon}(y - w_d),$$

is indicating a field that points in the same direction but decreases linearly. At $y = w_d$, it disappears. The voltage U_D, across the entire depletion zone, is obtained by integrating the field strength over y:

$$U_D = \frac{e}{2\varepsilon} \left(w_a^2 n_a + w_d^2 n_d \right)$$

is the result. The absolute value of U_D is not determined by the width of the depletion zone. It results from a thermodynamic equilibrium that weakly depends (logarithmically) on the concentration of the impurity atoms. For an optimisation of the technology parameters, it is therefore useful to to determine the total width $w = w_a + w_d$ as a function of the concentrations and U_D. One gets[2]

$$w = \sqrt{\frac{2 U_D \varepsilon}{e} \left(\frac{1}{n_a} + \frac{1}{n_d} \right)},$$

which shows that a low concentration obtains the key to a large depletion zone of impurity atoms.

At the same time, it becomes clear why the highest concentrations of semiconductor impurities can be found in microprocessors. They need a high transistor density and, therefore, narrow depletion zones.

4.3 The Equations of Laplace and Poisson

If the determination of electric fields by prior calculation of the potential Φ is to be advantageous, the first thing that needs to be clarified is how this potential can be determined. The two scientists, Laplace and Poisson, provided answers to this. The equations named after them will be presented below, and a sample application shall demonstrate their use.

---

2   Here, $w_a n_a = w_d n_d$ is used. This means that the donor side donates as many electrons as the acceptor side accepts.

Charge distributions determine fields

The following analysis will examine how the potential can be determined for a given distribution of free charges ρ_{free}: Without further restrictions, ◘ Eq. (4.6) and $\nabla \cdot (\varepsilon E) = \rho_{\text{free}}$ lead to the equation

$$
\begin{aligned}
E &= -\nabla\Phi &&|\varepsilon\cdot \\
\rightarrow \varepsilon E &= -\varepsilon\nabla\Phi &&|\nabla\cdot \\
\rightarrow \nabla \cdot (\varepsilon E) &= -\nabla \cdot (\varepsilon\nabla\Phi) &&|Eq.\ (3.12.1) \\
\rightarrow -\rho_{\text{free}} &= \nabla \cdot (\varepsilon\nabla\Phi)\,.
\end{aligned}
$$

With the help of the chain rule[3] the *general potential equation* for electric fields in matter,

$$-\rho_{\text{free}} = (\nabla\varepsilon) \cdot (\nabla\Phi) + \varepsilon\nabla^2\Phi\,, \tag{4.8}$$

is obtained. When integrating this equation, it may be necessary to ensure that ε is treated as a location-dependent integrand, not a prefactor.

If the potential is to be calculated in a homogeneous medium, then $\nabla\varepsilon = 0$, and the general equation simplifies to

$$-\rho_{\text{free}} = \varepsilon\nabla^2\Phi = \varepsilon\Delta\Phi\,, \text{ the *Poisson equation,*} \tag{4.9}$$

in which ε occurs as a constant. In the absence of matter, the choices are $\rho = \rho_{\text{free}}$ and $\varepsilon_0 = \varepsilon$.

If no charges are present,

$$0 = \nabla^2\Phi = \Delta\Phi\,, \text{ the *Laplace equation.*} \tag{4.10}$$

remains.

The solution of one of the last three equations determines the potential from a given charge distribution and, thus, indirectly, the electric field.

Spherical symmetry simplifies equations: sphere and point charge

The simplest example of using the Poisson and Laplace equations is the determination of the field of a homogeneously charged sphere in a vacuum. Even

---

3   Here: $\nabla \cdot (f\mathbf{a}) = f\nabla \cdot \mathbf{a} + (\nabla f) \cdot \mathbf{a}$.

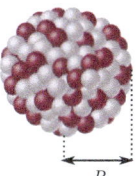

◘ Fig. 4.6 Nature only knows a few examples of homogeneously charged spheres. The Nuclei of heavy elements (here: uranium-235) are approximately such spheres

B

if there aren't too many examples of charged marbles, globes, or spheres, their analysis is helpful in several respects: The field from a point-like charge can be obtained by letting the radius of the sphere tend to zero. Formulas for arbitrary continuous charge distributions make use of this knowledge. Hence, an analysis of spheres paves the way away from differential equations towards integrals.

Large atomic nuclei like that of uranium (◘ Fig. 4.6) are approximately homogeneously charged spheres. Within the sphere ($r \leq R$) the Poisson equation applies, outside ($r > R$) it is the Laplace equation.

The charge distribution is spherically symmetric, so a spherically symmetric Potential $\Phi = \Phi(r)$ is expected. Consequently, $\partial\Phi/\partial\theta = \partial\Phi/\partial\varphi = 0$ may be assumed. The Poisson equation in spherical coordinates is then reduced to

$$\varepsilon\nabla^2\Phi_{\text{inner}}(r) = \frac{\varepsilon}{r^2}\frac{\partial}{\partial r}\left(r^2\frac{\partial\Phi_{\text{inner}}}{\partial r}\right) = -\rho_{\text{inner}}$$

where in this case, the partial derivatives ($\partial...$) can be replaced by (d...) because r is the only variabe.

Multiplication by $r^2 dr$ and integration then gives

$$\varepsilon r^2\frac{d\Phi_{\text{inner}}}{dr} = -\frac{1}{3}\rho_{\text{inner}}r^3 + c_1$$

with c_1 being the first constant of integration. This equation can also be solved by integration using variable separation ($\cdot dr/r^2$):

$$\varepsilon\Phi_{\text{inner}} = -\frac{1}{6}\rho_{\text{inner}}r^2 - \frac{c_1}{r} + c_2$$

Outside of the sphere, the Laplace equation is used. So

$$\varepsilon_0 \nabla^2 \Phi_{outer}(r) = \frac{\varepsilon_0}{r^2} \frac{\partial}{\partial r} \left(r^2 \frac{\partial \Phi_{outer}}{\partial r} \right) = 0$$

needs to be solved. Here, the use of ε_0 instead of ε indicates that there is no matter outside of the sphere. The solution to this equation can either be be calculated or copied from the solution of the Poisson equation with the replacements $\rho \to 0$ and $\varepsilon \to \varepsilon_0$:

$$\varepsilon_0 \Phi_{outer} = -\frac{c_3}{r} + c_4$$

The integration constants may be obtained are as follows:
- The potential must be finite at $r = 0$. Therefore $c_1 = 0$.
- The potential should disappear for large distances. As a consequence $c_4 = 0$. This is a choice-not necessary, but practical.
- For the field strength on the surface of the sphere to remain finite, the potential must be continuous. Therefore $\Phi_{inner}(R) = \Phi_{outer}(R)$ must be fulfilled, and as consequence

$$c_3 = \frac{\varepsilon_0}{\varepsilon} \left(\frac{\rho_{inner} R^3}{6} - c_2 R \right)$$

applies.
- According to Gauss' theorem for the electric field (3.13), at a material interface, the field strength of the vertical component must become smaller at the same rate as ε_r increases (see ◻ Eq. (3.11) and refraction laws for electric field lines, ◻ Fig. 3.8). Applied to the sphere,

$$\varepsilon E_{inner} = \varepsilon_0 E_{outer} \to \varepsilon \frac{d\Phi_{inner}}{dr} = \varepsilon_0 \frac{d\Phi_{outer}}{dr}$$

applies at $r = R$ with the consequence

$$c_3 = -\frac{\rho_{inner} R^3}{3} \, .$$

■ Finally, the solution for c_2 is

$$c_2 = \frac{\rho_{\text{inner}} R^2}{6} \left(1 + \frac{2\varepsilon}{\varepsilon_0}\right).$$

Altogether, the field strength can be written as

$$\Phi_{\text{inner}} = \frac{\rho_{\text{inner}}}{6\varepsilon_0\varepsilon_r} \left[R^2 (1 + 2\varepsilon_r) - r^2\right]$$

$$\Phi_{\text{outer}} = \frac{\rho_{\text{inner}} R^3}{3\varepsilon_0 r}. \tag{4.11}$$

The field inside the sphere is weakened to the extent ε_r grows. The potential becomes constant in the limit $\varepsilon_r \to \infty$, and the field strength inside the sphere tends to zero. Using $\boldsymbol{E} = -\nabla\Phi$, or $E = -\partial\Phi/\partial r$ the field strength progression can be determined as shown in ■ Fig. 4.7. The corresponding formulas for E are:

$$E_{\text{inner}} = \frac{\rho_{\text{inner}} r}{3\varepsilon_0\varepsilon_r}$$

$$E_{\text{outer}} = \frac{\rho_{\text{inner}} R^3}{3\varepsilon_0 r^2}. \tag{4.12}$$

The formula for the outer part of the electric field can be used to derive an expression for the field from a point-like charge. Using $Q = \rho_{\text{inner}} \cdot (4\pi R^3/3)$, one gets

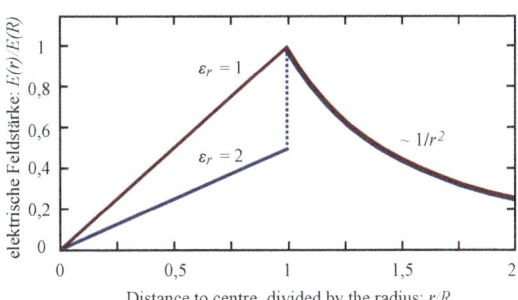

■ **Fig. 4.7** Electric field strength of a homogeneously charged sphere with radius R in a vacuum, normalised to its maximum value $E(R)$. If the relative dielectric constant of the sphere's material ε_r is not equal to one, the progression of the field is not continuous at $r = R$

$$\Phi_{\text{point-like charge}} = \frac{Q}{4\pi\varepsilon_0 r} \tag{4.13}$$

Also, one can use ◻ Eq. 4.11 to determine the potential curve and the field of a metal ball with radius R. Outside the sphere, the field and the potential must be the same as the ones of a point charge. There is no field within the sphere, so the potential must be constant $\Phi = Q/(4\pi\varepsilon_0 R)$. Of course, these results coincide with those derived from Gau's theorem for the electric field (3.14).

4

4.4 Calculating Potentials

Until the advent of computer-aided methods for calculating electrical potentials, strategies for solving the equations of Laplace and Poisson were key subjects of electrodynamics. In this chapter, three cases will be discussed. The simplest one occurs if all fields slowly disappear at large distances. A counterexample is a point-shaped charge carrier placed in front of a grounded conductor. For a limited set of geometries, this method allows an effective potential determination by using "mirror charges". Finally, it is indicated how the general case could be solved with the help of Green's functions.

Potentials for charge distributions follow from those of point-like charge carriers

The potential of a system with n point charges is (see ◻ Problem 4.1)

$$\Phi(\boldsymbol{r}) = \frac{1}{4\pi\varepsilon_0} \sum_{i=1}^{n} \frac{Q_i}{|\boldsymbol{r} - \boldsymbol{r}_i|} \tag{4.14}$$

and can be generalised to continuous charge distributions,

$$\Phi(\boldsymbol{r}) = \frac{1}{4\pi\varepsilon_0} \int_V \frac{\rho(\boldsymbol{r}')}{|\boldsymbol{r} - \boldsymbol{r}'|} \mathrm{d}V', \tag{4.15}$$

where the apostrophes ($'$) are intended to indicate the quantities to be integrated. The right part of this equation is called *Coulomb integral*. If the potential disappears at large distances, it allows us to determine the potential by

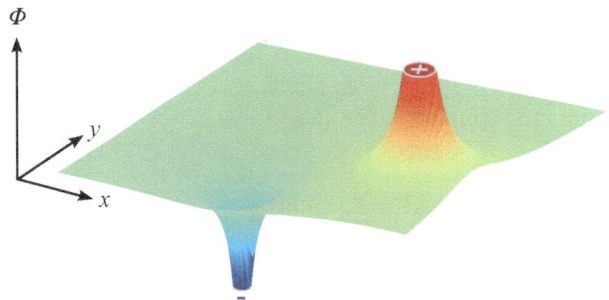

□ Fig. 4.8 Potential distribution of two charge carriers made of conductive material placed in the xy plane. The positively charged carrier is shown in red. The volcano-like shape is characteristic of charges far apart from each other

integration from a given charge distribution. Figure 4.8 shows the shape of the potential for one negatively charged and one positively charged conductive metal sphere calculated according to □ Eq. 4.15. If the charge carriers are placed in a medium with $\varepsilon_r \neq 1$, ε_0 in □ Eq. (4.15) has to be replaced by $\varepsilon = \varepsilon_0 \varepsilon_r$. In this case, the charges are then those of the carrier and not those bound in the medium. □ Equation (4.15) is the ideal starting point for the numerical calculation of potentials. Despite its seemingly simple form, an analytical computation of the integral is usually quite complicated. Even calculating the potential of a sphere (see e.g. [2]) is significantly more complex than the explicit solution of the Poisson and Laplace equations.

For the determination of the potential energy of a given charge distribution, it is helpful to rewrite □ Eq. (4.3) in such a manner that $\Phi(r)$ can be inserted. The result

$$W_{\text{pot}} = \frac{1}{4\pi\varepsilon_0} \sum_{i=1}^{n} \left(\sum_{k=i+1}^{n} \frac{Q_i Q_k}{r_{ik}} \right) = \sum_{i=1}^{n} Q_i \frac{1}{2} \left(\frac{1}{4\pi\varepsilon_0} \sum_{k=1}^{n} \frac{(1 - \delta_{ik}) Q_k}{|r_i - r_k|} \right)$$

contains a factor of $1/2$, which accounts for the fact that the second sum has more than twice as many terms as the one starting at $i + 1$. The Kronecker-δ_{ik},[4] which filters out the terms with $i = k$, makes the factor $1/2$ become exact.

---

4   It is $\delta_{ik} = 1$ for $i = k$ and otherwise equals zero.

In the transition to continuous distributions ($n \to \infty$), δ_{ik} may be omitted because the number of terms in which it is equal to one, grows proportionally to n, while the number at which it is equal to zero and grows in proportion to n^2. Therefore, $(1 - \delta_{ik}) \to 1$. The sum over k then turns into an expression for the potential similar to ◘ Eq. (4.14).

$$\text{for large } n: \quad W_{\text{pot}} \approx \frac{1}{2} \sum_{i=1}^{n} Q_i \Phi(r),$$

4

This result can be generalised to continuous distributions. The result,

$$W_{\text{pot}} = \frac{1}{2} \int_V \rho(r)\Phi(r)\mathrm{d}V, \tag{4.16}$$

is then an exact formula.

Grounded conductors in front of charges simulate mirror charges

Fields of charge carriers near grounded conductors of simple geometry can be calculated with the help of the following consideration: If the question "Can one find an additional charge carrier instead of the conductor so that the potential at the position of the conductor surface is equal to zero?" is answered by "yes", *mirror charges* are a good choice.

To determine the simplest case, we turn the argument around and look for the area between two point-shaped carriers of equal, opposite charges, where $\Phi_{\text{total}} = 0$. The symmetry of the problem suggests positioning the charge carriers at $r_1 = (-a, 0, 0)$ and $r_2 = (a, 0, 0)$. Then,

$$\Phi_{\text{total}} = \frac{-Q}{\sqrt{(x_0 + a)^2 + y_0^2 + z_0^2}} + \frac{Q}{\sqrt{(x_0 - a)^2 + y_0^2 + z_0^2}} = 0$$
$$\to x_0 = 0$$

must be fulfilled, while y_0 and z_0 can take arbitrary values. This set of conditions defines the (flat) yz plane. Consequently, the electric field of a point-like charge carrier at a distance a to a grounded conductive plane is the same as that of two oppositely charged carriers whose distance is $2a$. It looks like the mirror image, as indicated in ◘ Fig. 4.9. The second charge is referred to as *mirror charge*.

□ **Fig. 4.9** Mirror charge: A point-shaped charge carrier in front of a grounded, flat conductor has the same field as a pair of oppositely charged carriers

But what happens if the two carriers have different charges? In this case, there is no obvious symmetry, and so a charge carrier is placed at the origin of the coordinate system: $r_0(Q_0) = (0, 0, 0)$ and $r_1(Q_1) = (b, 0, 0)$. The potential is then zero wherever

$$\Phi_{\text{total}} = \frac{Q_0}{\sqrt{x_0^2 + y_0^2 + z_0^2}} + \frac{Q_1}{\sqrt{(x_0 - b)^2 + y_0^2 + z_0^2}} = 0$$

$$\rightarrow x_0^2 + y_0^2 + z_0^2 = \frac{-b^2}{1 - \left(\frac{Q_1}{Q_0}\right)^2} + \frac{2b}{1 - \left(\frac{Q_1}{Q_0}\right)^2} \cdot x_0$$

is fulfilled. The latter equation describes a sphere displaced along the x-axis. This can be shown as follows: Let R be the radius of a sphere and v the distance of the centre to the point $(0, 0, 0)$, then the corresponding equation

$$(x_0 - v)^2 + y_0^2 + z_0^2 = R^2$$

$$\rightarrow x_0^2 + y_0^2 + z_0^2 = -(v^2 - R^2) + 2vx_0$$

has the same form as that for the condition for $\Phi_{\text{total}}(x_0, y_0, z_0) = 0$. A coefficient comparison provides the distance and radius of the sphere: $v = bQ_0^2/(Q_0^2 - Q_1^2)$ and $R = |bQ_0Q_1/(Q_0^2 - Q_1^2)|$.

For a compact formulation of the result, it is advisable to place the sphere in the centre [3]. The smaller charge is then within the sphere, as shown in □ Fig. 4.10. The complete description is then[5]

---

5   Here, $r_m = v$ and $r_p = vb$ are used for the transormation.

■ **Fig. 4.10** The field of a
charge carrier near a grounded
sphere has the same shape as
one with a carrier of opposite
charge inside the sphere. If the
real charge carrier is outside
the sphere, the blue field lines
indicate a real field, the red
ones indicate a fictitious field

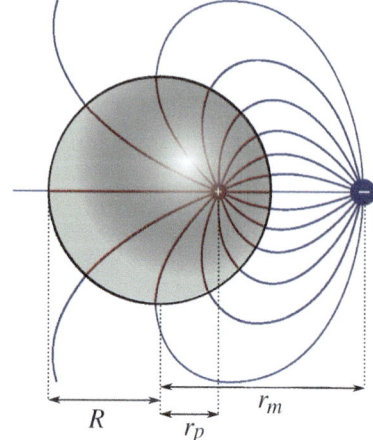

$$r_p r_m = R^2 \;\; \text{and} \;\; -\frac{Q_p}{Q_m} = \sqrt{\frac{r_p}{r_m}} \;\; \text{(mirror charge, sphere)}.$$

These formulas are independent of whether the real charge carrier is outside
or inside the sphere.

Since two charges can only be of the same or different size, the above
results have the following implication:

❯ Limitations for Mirror Charges

The field of a point-shaped charge carrier near a grounded conductor can be
determined using mirror charges if, and only if, the conductor has the shape
of a sphere or its limit of a flat surface.

Green's function helps if boundary conditions are non-trivial

So far, it has been assumed that the potential Φ vanishes at large distances.
Calculating the integral in ■ Eq. (4.15) then suffices to determine the potential at any location. However, the example of mirror charges shows that there
are Applications where either the values for Φ or their derivatives are
specified at given boundaries.

If the values for a potential are given at the boundary, then the situation
is referred to as a *Dirichlet problem*. If the derivatives are specified, *Neumann
problem* is the term. The underlying formalism will be presented below:

George Green found that even under such conditions the potential can be determined via integration, provided that a function G (like "*Green's function*") can be found that fulfils

$$\Phi(\boldsymbol{r}) = \frac{1}{\varepsilon} \int\limits_V \rho(\boldsymbol{r}') \frac{1}{4\pi|\boldsymbol{r} - \boldsymbol{r}'|} \mathrm{d}V' \quad | \text{ if } \Phi \to 0 \text{ for } r \to \infty$$

$$\Phi(\boldsymbol{r}) = \frac{1}{\varepsilon} \int\limits_V \rho(\boldsymbol{r}') G(\boldsymbol{r}, \boldsymbol{r}') \mathrm{d}V' \quad | \text{ any boundary conditions.}$$
(4.17)

It may be surprising at first sight, but Green's function is simply the solution of the Poisson equation for a point-like charge for the boundary conditions given. For a formal investigation of this statement, the mathematical description of a point-like distribution is needed. The δ distribution shown in ◻ Fig. 4.11 offers just that. This distribution is implicitly defined by replacing an integral with its integrand at a point

$$\int\limits_{-\infty}^{\infty} f(x')\delta(x - x')\mathrm{d}x' = f(x)$$

defined by the δ distribution. Applied to charge distributions, this means, for example

◻ **Fig. 4.11** Graphical approximation to the δ distribution. Imagine an area of size 1 in the $x - f(x)$-plan which becomes increasingly narrower and higher. The δ distribution is approached to the extent that the width of the area approaches zero (and the height becomes very large)

$$\int\limits_{\text{entire universe}} \rho(\mathbf{r}')\mathrm{d}V' = Q(\mathbf{r}) \quad \text{if} \quad \rho(\mathbf{r}') = Q\delta^3(\mathbf{r} - \mathbf{r}') \tag{4.18}$$

with $\delta^3(\mathbf{r} - \mathbf{r}') = \delta(x - x')\delta(y - y')\delta(z - z')$.

While replacing a sum with an integral allows the transition from a discrete to continuous distributions, the δ distribution paves the way back. The δ distribution allows us to write down the Poisson equation for a distribution concentrated at a single point $\mathbf{r}$:

$$\Delta G(\mathbf{r}, \mathbf{r}') = -\delta^3(\mathbf{r} - \mathbf{r}') \quad |\text{for a point } \mathbf{r} \text{ while integrating over } \mathbf{r}' \tag{4.19}$$

The formal detail should be noted that Δ acts on $\mathbf{r}$, only - not on $\mathbf{r}'$.

It turns out that G is the Green's function we seek. Because taking the second derivative in ◘ Eq. (4.17) for charges in a neither moving nor changing volume V gives

$$\begin{aligned}
\Delta\Phi(\mathbf{r}) &= \frac{1}{\varepsilon}\int\limits_V \rho(\mathbf{r}')\left[\Delta G(\mathbf{r}, \mathbf{r}')\right]\mathrm{d}V' \\
&= \frac{-1}{\varepsilon}\int\limits_V \rho(\mathbf{r}')\delta^3(\mathbf{r} - \mathbf{r}')\mathrm{d}V' \\
&= \frac{-1}{\varepsilon}\rho(\mathbf{r}) ,
\end{aligned}$$

which is again the Poisson equation. Therefore, one can conclude:

The potential $\Phi(\mathbf{r})$ associated with a charge distribution $\rho(\mathbf{r})$ can be determined by the integral

$$\Phi(\mathbf{r}) = \frac{1}{\varepsilon}\int\limits_V \rho(\mathbf{r}')G(\mathbf{r}, \mathbf{r}')\mathrm{d}V' \tag{4.20}$$

assuming Green's function $G(\mathbf{r}, \mathbf{r}')$ is a solution of the equation

$$\Delta G(\mathbf{r}, \mathbf{r}') = -\delta^3(\mathbf{r} - \mathbf{r}'), \tag{4.21}$$

where Δ acts on $\mathbf{r}$ and not on $\mathbf{r}'$.

If the potential disappears at infinity, Green's function is already known from ◻ Eq. (4.17):

$$G(r, r') = \frac{1}{4\pi|r - r'|} .$$

For other boundary conditions, it is best to write it in the form

$$G(r, r') = \frac{1}{4\pi|r - r'|} + F(r, r') , \qquad (4.22)$$

because then the second part, $F(r, r')$, must fulfil the Laplace equation (see also ◻ Problem 4.8).

The simplest example, $F(r, r') = 0$ and $\rho = Q\delta^3(r - r')$ may serve as a "confidence-raising measure". Inserting these conditions into ◻ Eq. (4.22) gives

$$\Phi(r) = \frac{1}{\varepsilon} \int_V \rho(r')G(r, r')\mathrm{d}V' = \frac{1}{\varepsilon} \int_V Q\delta^3(r - r')\frac{1}{4\pi|r - r'|}\mathrm{d}V' = \frac{Q}{4\pi\varepsilon|r|} ,$$

i.e. the known formula for a point charge.

Overall, field calculations using Green's function follow the general principle:

> The potential of a set of charges is simply the sum of their potentials. Consequently, the potential of an arbitrary charge distribution is known if the potential of a point-like charge is known. It results from folding the potential of the point-like charge with the charge distribution.

The determination of Green's function under arbitrary boundary conditions is beyond the scope of this book. The literature (see [4, 5, 6]) describes many solving strategies.

Unfortunately, Green's function in different books have different normalisations. Factors of 2, π and ε vary, depending on the chosen system of units and depending on which of factors occuring in ◻ Eq. (4.17) are included in Green's function. In this book, the choice is the simplest differential ◻ Eq. (4.21).

4.5 Electric Multipoles

The exact calculation of electric fields is often complex or even impossible. Therefore, approximation methods for various recurring arrangements have been developed. For electrically neutral arrangements (i.e. those with the same amount of positive and negative charge), the *multipole expansion* is the method of choice. This method orders its approximation terms according to their field contribution at large distances.

This section first describes the simplest and often most important multipole, the *dipole*. Its more detailed investigation leads to a physical interpretation of the vector "polarization", **P**. It will be the fraction of the electric field that molecular dipoles contribute. Then, higher-order multipoles will be discussed.

Electric dipoles respond to fields even though they are neutral

The simplest form of a dipole is as shown in ◘ Fig. 4.12. It consists of two point-like charge carriers at a distance **d** from each other. It is characterised by its *dipole moment* **p** $= Q$**d** which is a measure of its internal charge separation. It grows both with the amount of the charges and the distance between their carriers. For a system of N point charges, the definition of the dipole moment is generalised as follows:

$$\boldsymbol{p} = \sum_{i=1}^{N} Q_i \boldsymbol{r}_i \tag{4.23}$$

◘ **Fig. 4.12** The simplest form of an electric dipole. It consists of two carriers of opposite charges at a distance **d**

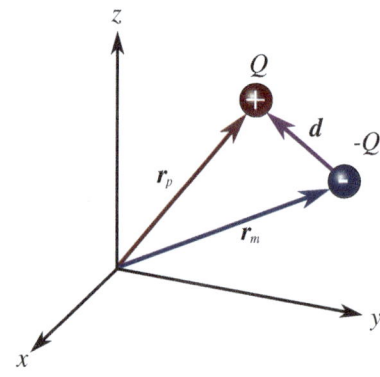

This definition assumes that the system is neutral. Because as stated in ◘ Problem 4.9, only then p is independent of the choice of the reference system. For continuous distributions,

$$p = \int \rho(r')r'\mathrm{d}V' \tag{4.24}$$

is the counterpart to ◘ Eq. (4.23).

Solids often contain a vast number of individual dipoles, each p_i consisting of just a few individual atoms. Solids may therefore more easily be characterised by the *dipole density* P

$$P = \left(\sum_{i=1}^{N} p_i\right)/V$$

with V being the volume.

The electric field imposes a torque on a dipole

In an electric field, negative charges are exposed to the opposite forces than positive charges are. So, every dipole experiences a torque τ, which only becomes zero when it aligns with the field, $p \parallel E$. Otherwise, the system is "stretched", only. By definition, a torque is the cross product of a force F and a distance vector r_{axis}. Applied to the dipole shown in ◘ Fig. 4.12, this means

$$
\begin{aligned}
\tau &= [d/2 \times (F(Q)] + [(-d/2) \times F(-Q)] \\
&= [d/2 \times (QE)] + [d/2 \times (QE)] \\
&= Qd \times E \\
\rightarrow \tau &= p \times E \,,
\end{aligned}
\tag{4.25}
$$

provided the origin of the coordinate system is placed midway between the charges. The last line is also valid if the dipole is composed of more than two individual charges. And its calculation does not require any knowledge of the centre of gravity of the charges. But the torque's physical meaning is linked to this centre. In ◘ Fig. 4.12), the torque acts around the axis of rotation which passes through the middle of the distance vector d.

Internal dipoles create the electric field $E_{bound} = -P/\varepsilon_0$

The most common notation of Gaus' theorem for the electric field in matter is $\rho_{free} = \nabla \cdot (\varepsilon_0 E + P)$. In this formula, two apparently completely different quantities are added to each other. This should be puzzling as it looks like adding "pears and apples": The sum of a field and a dipole density, trimmed for consistent units by inserting a factor ε_0? The following argument will show why the summation is possible and even correct: The polarised molecules generate a field $E_{bound} = -P/\varepsilon_0$. So, what looks like pears and apples is nothing but the difference between the entire field E and one of its contributions (or shares).

For a better understanding, consider the part of a solid shown in ◘ Fig. 4.13. The cube of edge length l contains a number N dipoles with dipole moments p_1, each. They are densely packed. All dipoles are aligned perpendicular to the top edge of the cube. This creates a configuration in which layers with positive and negative charges alternate. Inside the cube, these layers neutralise each other. An excess of charges remains on the top and bottom layers. Therefore, the following equations are fulfilled:

$$
\begin{array}{ll}
p_1 = Q_1 \cdot d & \text{| charge and length of a single dipole} \\
d = l/N^{1/3} & \text{| the number of dipoles in a row is } N^{1/3} \\
Q = Q_1 \cdot N^{2/3} & \text{| excess of charge at the surface} \\
E_{dipoles} = -Q/(\varepsilon_0 l^2) & \text{| field from surface charges}(\leftarrow \text{capacitor})
\end{array}
$$

Now, the electric field can be written as a function of the individual dipoles p_1. The result

$$
E_{dipoles} = \frac{-p_1 N}{\varepsilon_0 l^3} = \frac{-p_1}{\varepsilon_0} \frac{N}{V} \tag{4.26}
$$

◘ **Fig. 4.13** Cube-shaped cutout made from a polarised solid. The coordinate system is chosen so that all dipoles point upwards (in the z-direction): $\mathbf{p}_1 = (0, 0, p_1)$. The charges $+Q$ and $-Q$ at the surface create an electric field $\mathbf{E} = (0, 0, -E)$

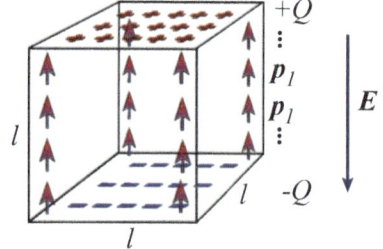

coincides with the one obtained by analysing Maxwell's equations (see ▣ Table 3.1) if the field generated by the dipoles is equated with the Field component E_{bound}. The name "dipole density" for $P = -E_{bound}/\varepsilon_0$ now has an obvious meaning (except for a historically motivated minus sign): It is the number of individual dipoles per volume multiplied by their dipole moments.

❯ Field Contribution due to Polarisation
The bound charges of a neutral solid are shifted against each other by the process of polarisation. This process creates internal dipoles. The electric field of these dipoles can be identified as $-P/\varepsilon_0$ in Maxwell's equations.

Note that Maxwell's equations do not contain the process of polarisation, but rather its result: an electrical field from molecules.[6]

❶ Model Limitations
Internal dipoles are not as tightly packed as assumed for the derivation of ▣ Eq. (4.26). Normally, p_1 is the product of a length considerably smaller than $l/N^{-1/3}$ and a correspondingly larger charge. If this were not the case, every polarisation would have to be accompanied by a massive deformation. The term "dipole density" is therefore based on idealised assumptions. In contrast, the term "field fraction from bound charges" is always correct.

The field of an electric dipole rapidly decreases with the distance

The electric field of a dipole is the sum of the fields of its charges. If the z-axis is chosen to be the vector from the negative charge to the positive charge, as shown in ▣ Fig. 4.14, the charges are at the locations $r(Q) = (0, 0, d/2)$ and $r(-Q) = (0, 0, -d/2)$. The field only depends on z and the Distance y to the z-axis

$$E_z = \frac{Q}{4\pi\varepsilon_0}\left[\frac{z - d/2}{\left|(y^2 + (z - d/2)^2)^{3/2}\right|} - \frac{z + d/2}{\left|(y^2 + (z + d/2)^2)^{3/2}\right|}\right]$$

$$E_y = \frac{Q}{4\pi\varepsilon_0}\left[\frac{y}{\left|(y^2 + (z - d/2)^2)^{3/2}\right|} - \frac{y}{\left|(y^2 + (z + d/2)^2)^{3/2}\right|}\right],$$

---

6   The vector P is often given the same name as the process: "polarisation". This may not be the best choice, as one usually distinguishes between a process and its result (f.ex. birth and baby).

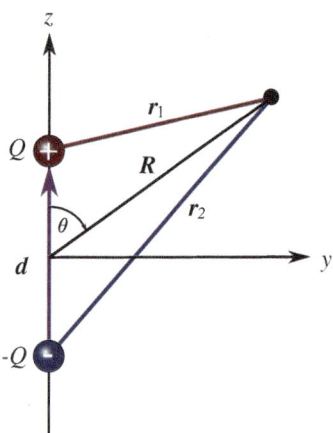

◻ **Fig. 4.14** Dipole in a coordinate system aligned to it. The quantities shown are used in the calculation of the field. The variable y is the distance from the dipole axis, it is always positive

and has the following properties: The z component of the field has the largest values between the charge carriers. Only there, the fields of the two charges are added to each other. For $z > 0$, the radial component points outwards ($E_y > 0$), for $z < 0$ inwards.

If one is interested in the field at a large distance, it is advantageous to calculate the potential first, perform a expansion to powers of $|a|/R$ of the result and finally, determine the field using $E = -\nabla\Phi$ [7]. For the dipole shown in ◻ Fig. 4.14, one gets

$$\Phi = \frac{Q}{4\pi\varepsilon_0}\left(\frac{1}{r_1} - \frac{1}{r_2}\right) \qquad \text{| potential, exact}$$

$$r_1 = |R - d/2| = \sqrt{R^2 + \frac{d^2}{4} - Rd\cos\theta} \quad \text{| distance to } +Q, \text{ exact}$$

$$\frac{1}{r_1} \approx \frac{1}{R}\left(1 + \frac{d\cos\theta}{2R}\right) \qquad \text{| because of } 1/\sqrt{1+\varepsilon} \approx 1 - \varepsilon/2$$

$$\frac{1}{r_2} \approx \frac{1}{R}\left(1 - \frac{d\cos\theta}{2R}\right) \qquad \text{| like } 1/r_1$$

$$\rightarrow \Phi \approx \frac{Q}{4\pi\varepsilon_0}\cdot\frac{d\cos\theta}{R^2} \qquad \text{| approximation for } d/R \ll 1 \qquad (4.27)$$

or in vector notation

$$\Phi \approx \frac{p \cdot R}{4\pi\varepsilon_0 R^3} \quad \text{(approximation for } a/R \ll 1). \tag{4.28}$$

The potential decreases proportionally to the square of the distance to the dipole. The field is then calculated as $E = -\nabla\Phi$, i.e.

$$
\begin{aligned}
-4\pi\varepsilon_0 E &\approx \nabla\left(\frac{p \cdot R}{R^3}\right) & | \quad \text{chainrule} \\
&= \left(\nabla\frac{1}{R^3}\right) p \cdot R + \frac{1}{R^3}\nabla(p \cdot R) \; | \quad \text{with } \nabla(p \cdot R) = p \\
&= -\frac{3}{R^5}R(p \cdot R) + \frac{1}{R^3}p
\end{aligned}
$$

and finally,

$$E \approx \frac{1}{4\pi\varepsilon_0 R^3}\left\{3R\left(\frac{p \cdot R}{R^2}\right) - p\right\} \quad \text{(approximation for } a/R \ll 1), \tag{4.29}$$

which shows that the field decreases in proportion to $1/R^3$, i.e. more rapidly than that of a single charge.

The same result can be obtained by calculating one cartesian component first, and then generalising the result into three dimensions:

$$
\begin{aligned}
E_x &\approx -\frac{1}{4\pi\varepsilon_0}\frac{\partial}{\partial x}\left(\frac{xp_x + yp_y + zp_z}{(x^2 + y^2 + z^2)^{3/2}}\right) \\
&= -\frac{1}{4\pi\varepsilon_0}\left\{\frac{p_x}{R^3} + p \cdot R \cdot \left(\frac{-3}{2}\right) \cdot \frac{1}{R^5} \cdot 2x\right\} \\
&= +\frac{1}{4\pi\varepsilon_0 R^3}\left\{3x\frac{p \cdot R}{R^2} - p_x\right\}
\end{aligned}
$$

This leads to the same result.

Multipole expansions give approximations for large distances

Multipole developments are used to find manageable expressions for the electric field, even if the charge distributions are complicated. A multipole expanded field is approaching the actual field if either the distance is much

4

greater than the size of the charge distribution or if many terms of the expansion are taken into account.

The archetypes of multipoles are shown in ◘ Fig. 4.15. One peculiarity is that for all configurations on the left of each multipole, the respective moment equal to zero. The dipole has a total charge (monopole) of zero, the quadrupole shown has both the total charge and the dipole moment zero. (...)

From a mathematical point of view, the multipole expansion is an approximation method, in which the series expansion (4.27) of the root function is carried out up to arbitrary powers of ε. The result is almost always an expression for the field which follows a simple power law at large distances - even in the case of complicated charge distributions. Consequently, multipole expansions are particularly helpful if distances to the charge distribution are much larger than the size of the charge distribution.

For a general description, the field at a point P in ◘ Fig. 4.16 shall be described in such a way that the result of the expansion quickly approaches the exact one as the distance increases. For this purpose, the distance $|R - r'|$ to a volume element dV' occurring in the Coulomb integral (4.15) shall be expanded around the distance vector R towards the centre:

$$\frac{1}{|R - r'|} = \frac{1}{\sqrt{R^2 + r'^2 - 2Rr'\cos\theta}} = \frac{1}{R\sqrt{1 + \epsilon}}$$

◘ **Fig. 4.15** Archetypes of multipoles

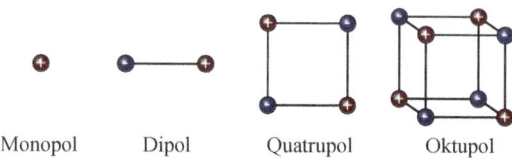

Monopol   Dipol   Quatrupol   Oktupol

◘ **Fig. 4.16** Variables used to calculate the electric potential at point **P** at a distance **R** from the center of an infinitesimal Volumes dV' existing body

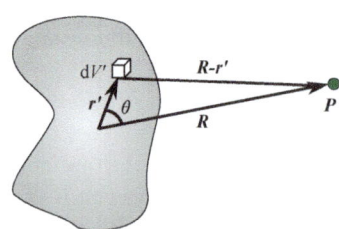

with

$$\epsilon = \left(\frac{r'}{R}\right)^2 - \left(\frac{r'}{R}\right) \cdot 2\cos\theta = \left(\frac{r'}{R}\right) \cdot \left(\frac{r'}{R} - 2\cos\theta\right).$$

The first terms of the series expansion are

$$\frac{1}{\sqrt{1+\epsilon}} \approx 1 - \frac{1}{2}\epsilon + \frac{3}{8}\epsilon^2 - \frac{5}{16}\epsilon^3 + \dots$$

and lead to the series expansion sorted by powers of r'

$$\frac{1}{|R-r'|} = \frac{1}{R}\left\{1 + \left(\frac{r'}{R}\right)\cos\theta + \left(\frac{r'}{R}\right)^2 \frac{3\cos^2\theta - 1}{2} + \left(\frac{r'}{R}\right)^3 \frac{5\cos^3\theta - 3\cos\theta}{2} + \dots\right\}$$

$$(4.30)$$

where the angle-dependent parts are called *Legendre polynomials* $P_i(\cos\theta)$.[7] Surprisingly (see e.g. [2]), if all powers of ϵ are taken so that the formula is extended to infinity,

$$\frac{1}{|R-r'|} = \frac{1}{R}\sum_{i=0}^{\infty}\left(\frac{r'}{R}\right)^i P_i(\cos\theta)$$

the expansion becomes an exact identity.

The terms may now be sorted by powers of r':

$$\Phi(R) = \sum_{i=0}^{\infty}\Phi_i(R) = \frac{1}{4\pi\varepsilon_0}\frac{1}{R}\sum_{i=0}^{\infty}\int_V\left(\frac{r'}{R}\right)^i P_i(\cos\theta)\rho(r')dV' \qquad (4.31)$$

The index i specifies the power of r'. The higher the index i, the faster the term diminishes with the distance. ◻ Equation (4.31) is know as the *multipole expansion* of the electrical potential. The discrete variant of ◻ Eq. (4.31),

---

7   In atomic physics, these polynomials also describe the angular dependence of the electron orbitals and so the "shape" of the atoms. That's why Wolfgang Pauli called them "Urformen der Natur" ($\approx$ archeformes of nature).

$$\Phi(\boldsymbol{R}) = \sum_{i=0}^{\infty} \Phi_i(\boldsymbol{R}) = \frac{1}{4\pi\varepsilon_0} \frac{1}{R} \sum_{i=0}^{\infty} \sum_{n=1}^{N} \left(\frac{r_n}{R}\right)^i Q_n P_i(\cos\theta_n) \qquad (4.32)$$

gives the expansion for N point-like charge carriers. A discussion of the connection between the general description of dipoles and quadrupoles and the terms used in in ◻ Eqs. (4.31, 4.32) can be found in [8].

Multipole Expansions and Power Laws

If the distance to a system of charge carriers is much greater than its size, the potential follows a power law, whose exponent is given by the first non-vanishing term of the multipole expansion.

The best approximations will be obtained if the origin of the coordinate system is placed into the centre of gravity of the charges. If possible, the calculation should be carried out in the plane of the charge carriers.

Four charges can either form a dipole or a quadrupole

◻ Figure 4.17 shows four charges placed in the yz plane. The potential at the point $\boldsymbol{P} = (0, 0, R)$ shall be determined as a function of the two angles θ, δ assuming the same distance r to the coordinate origin for all charges.

◻ **Fig. 4.17** Four charges in the yz plane at the same distance from the origin, and at the same distance to the straight line shown. The line is inclined by an angle θ. Depending on the choice for the signs of the charges, they form a monopole, a dipole, or a quadrupole

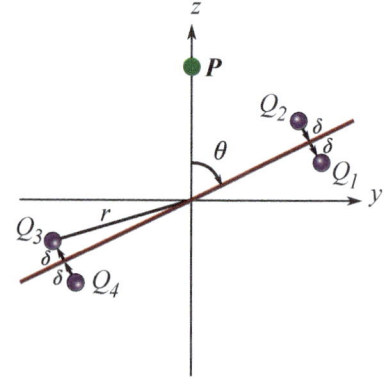

In this example $\theta_1 = \theta + \delta$, $\theta_2 = \theta - \delta$, $\theta_3 = \theta + \pi + \delta$ and $\theta_4 = \theta + \pi - \delta$. A dipole emerges, if $Q_1 = Q_2 = Q$ and $Q_3 = Q_4 = -Q$ is chosen. One then obtains

for $Q_1 = Q_2 = -Q3 = -Q_4$:

$$\Phi_0(\boldsymbol{P}) = 0 \qquad\qquad\qquad \text{(total charge 0)}$$

$$\Phi_1(\boldsymbol{P}) = \frac{Qr}{\pi\varepsilon_0 R^2}\cos\theta\cos\delta \qquad \text{(dipole term)}$$

$$\Phi_2(\boldsymbol{P}) = 0 \qquad\qquad\qquad \text{(no quadrupole term)},$$

i.e. a dipole. The larger the angular distance 2δ of the charge carriers, the smaller the value for the potential becomes. The potential terms Φ_0, Φ_1 and Φ_2, determine the values for the potential at the point $\boldsymbol{P} = (0, 0, R)$. For $\delta = 0$, the result corresponds to ◾ Eq. (4.28). For $\delta = \pi/2$, charges of the same sign are precisely opposite. In this case, the dipole moment vanishes.

An alternating charge distribution around the origin gives

for $Q_1 = -Q_2 = Q3 = -Q_4$:

$$\Phi_0(\boldsymbol{P}) = 0 \qquad\qquad\qquad\qquad \text{(total charge 0)}$$

$$\Phi_1(\boldsymbol{P}) = 0 \qquad\qquad\qquad\qquad \text{(no dipole term)}$$

$$\Phi_2(\boldsymbol{P}) = -\frac{3Qr^2}{4\pi\varepsilon_0 R^3}\sin(2\theta)\sin(2\delta) \quad \text{(quadrupole term)},$$

i.e. a quadrupole term with its maximum when the charge carriers form a square, that means if $\delta = \pi/4$. For $\delta = 0$, $\delta = \pi/2$, ... the charges with different signs are placed exactly "on top of each other". Just like in the case of the dipole, the angle δ changes the overall strength of the potential but not its angular dependence. The term $\sin(2\delta)$ does not affect the shape of the field.

A change of perspective allows to calculate the field in the entire plane

If the multipole terms Φ_1, Φ_2, ... are known, they can be used to calculate the potential and, subsequently, the electric field. The result is a good approximation if the distance R is much larger than that charge arrangement.

First, the potential must be determined for a point not necessarily lying on the z-axis. For this purpose, one imagines the charges shown in ◾ Fig. 4.17 to be rotated backwards by the angle θ. Then, $\boldsymbol{R} = R(0, -\sin\theta, \cos\theta)$ and the

4

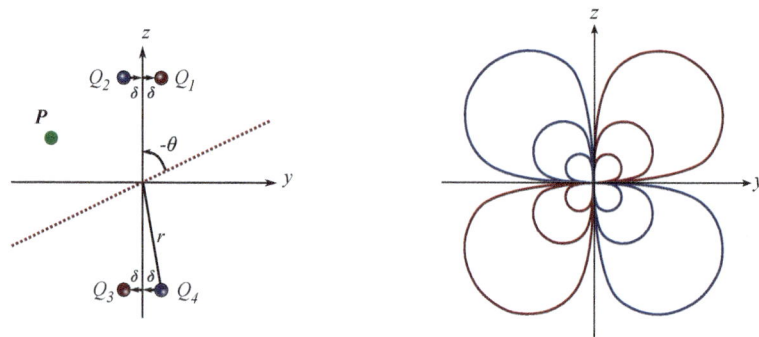

◻ Fig. 4.18   Left: the configuration shown in ◻ Fig. 4.17, rotated by the angle $-\theta$ and realized as a quadrupole. Right: equipotential lines for this quadrupole. The values for Φ_2 vary in proportions from 1 (inside) to 0.1 to 0.01 (outside). Positive potentials are shown in red, and negative ones are shown in blue. The units of the lengths y, z are arbitrary

charge configuration is aligned with the z-axis, as shown in ◻ Fig. 4.18. A rotation by the angle θ now means rotating the position of the observation point P. If θ takes any value from 0 to 2π, the potential can be determined at each point. Formally, the change in perspective means that θ ("from the fixed point to the arbitrarily oriented dipole") is replaced by $-\theta$ ("from the fixed dipole to the arbitrarily oriented point").

For a certain value of Φ, the multipole terms calculated above establish a connection between the distance R and the angle θ. The equipotential lines in the yz plane defined in this way are shown in ◻ Fig. 4.18. The drop in the quadrupole potential with the distance from the centre is highlighted by the fact that increasing the distance by a factor of 4.64 corresponds to a drop of the potential by a factor of one hundred.

The electric field can now be calculated as the gradient of the potential: For the first non-vanishing term Φ_1 of the dipole one obtains for example

$$E_{\text{dipol}} = -\nabla\Phi_1(-\theta) = -\nabla\frac{Qr}{\pi\varepsilon_0 R^2}\cos\theta\sin\delta$$

$$\rightarrow E_{\text{dipole},R} = -\frac{\partial}{\partial R}\Phi_1 \quad = \frac{2Qr}{\pi\varepsilon_0 R^3}\cos\theta\sin\delta$$

$$\rightarrow E_{\text{dipol},\theta} = -\frac{1}{R}\frac{\partial}{\partial\theta}\Phi_1 \quad = \frac{Qr}{\pi\varepsilon_0 R^3}\sin\theta\sin\delta$$

with θ being the angle to the z-axis, assuming the dipole is aligned parallel to it. The calculation of the quadrupole field can be done analogously.

The multipole expansion can be extended to three dimensions. This is widely discussed in the literature (e.g. [4]).

4.6 **Capacitors**

In engineering, the capacitance C is usually understood as the property a capacitor, to keep two charges $+Q$ and $-Q$ separated if a voltage U is applied. There is a charge on the anode

$$Q = CU. \tag{4.33}$$

On the cathode, it is $Q = -CU$ at the same time.

In a transmission line, the charges of both wires contribute to the field...

The capacitance of a twin cable shall be determined first. The field of the left conductor shown in ◘ Fig. 4.19 is cylindrically symmetrical. At a distance of R from the centre of the wire, it has a strength of $E_1 = Q/(2\pi R\, l\varepsilon)$. To calculate the voltage, the direct path from one surface to the opposite one is used for integration

$$U_1 = \int_{r}^{d-r} E_1 \mathrm{d}x' = \frac{Q}{2\pi l\varepsilon}\ln\left(\frac{d-r}{r}\right)$$

◘ **Fig. 4.19** Two parallel, oppositely charged wires of length l with radii a. The distance between the centre axes is d

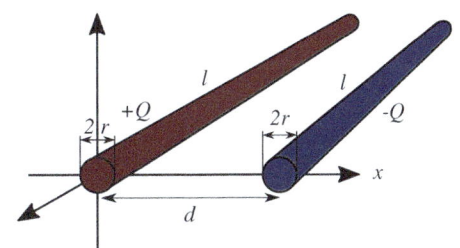

The total voltage must be twice as large as U_1 because the field strength of the right wire points in the same direction. On average, it is equally strong. Consequently, the capacitance is

$$C_{\text{twin line, thin}} = \frac{Q}{U} = \frac{Q}{2U_1} = \frac{\pi l \varepsilon}{\ln\left(\frac{d-r}{r}\right)}. \tag{4.34}$$

John David Jackson [9] noticed in the 1970s that the equipotential lines of a thin pair of wires form cylinders slightly shifted outwards. This insight led to a formula for the capacitance of thick wires. If the wires are placed in a way that their surfaces coincide with the equipotential lines of a thin wire, the formula for thick cables can be deduced from the one for thin ones. Unfortunately, the result

$$C_{\text{twin line, general}} = \frac{\pi \varepsilon l}{\text{arcosh}\left(\frac{d}{r}\right)}$$

requires a considerable amount of prior calculations.

..., in a spherical capacitor, they don't.

If a charged sphere with radius R_1 is placed inside a larger hollow sphere of radius R_2, the result is a spherical capacitor. Since no field is inside a charged hollow sphere, its potential must be constant. Within the sphere, it is $\Phi_2 = Q_2/(4\pi\varepsilon_0 R_2)$ if the potential is assumed to approach zero at large distances. The field between the two spheres is determined exclusively by the charge Q_1 of the inner sphere, as may be proven by using Gauss' law. The potential difference, i.e. the voltage between the two spherical surfaces is

$$U = \Phi_1 - \Phi_2 = \frac{Q_1}{4\pi\varepsilon_0 R_1} - \frac{Q_1}{4\pi\varepsilon_0 R_2} = \frac{Q_1}{4\pi\varepsilon_0}\left(\frac{1}{R_1} - \frac{1}{R_2}\right)$$

The field outside of the outer sphere disappears, as made plausible in ◻ Fig. 4.20, for $Q_2 = -Q_1$. If these charges are chosen, the capacitance is

$$C_{\text{spherical capacitor}} = \frac{Q_1}{U} = \frac{4\pi\varepsilon_0}{\frac{1}{R_1} - \frac{1}{R_2}} \tag{4.35}$$

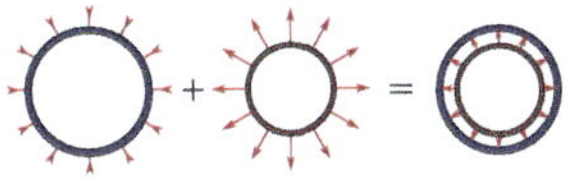

□ **Fig. 4.20** Field of a spherical capacitor as obtained by
a superposition of the fields
of a negatively charged hollow sphere and a positively
charged smaller sphere

and has an interesting property in the limit $R_2 \rightarrow \infty$, in which all charged
objects are much further away than R_1,

$$C_{\text{sphere}} = 4\pi\varepsilon_0 R_1 .$$

Consequently, one can allocate a capacitance to chargeable individual objects
provided that all other electrically relevant objects are sufficiently far away.
This capacitance indicates which potential change corresponds to which
transfer of charge. The magnitude of such capacities is exemplified in
□ Problem 4.4.

The spherical capacitor has another, technically much more relevant
limiting case: Imagine both spheres inflated to astronomical sizes but still
similarly large. In this limit,

$$C_{\text{spherical capacitor}} \approx \frac{\varepsilon A_{\text{sphere}}}{d}$$

with $d = R_2 - R_1$ is obtained. The spheres could be cut into many small
individual fragments to obtain pairs of electrodes that can no longer be distinguished from a plate capacitor with the electrode spacing d. The well-known
result for each fragment

$$C_{\text{plate capacitor}} = \frac{\varepsilon A}{d}$$

can be formulated like this: "The field of a plate capacitor can be calculated
from the charge of only one of the two electrodes because the counter
electrode is the extreme case $R \rightarrow \infty$ of a sphere with no field in its inside."

Note that the cable capacitor is not a limiting case of the plate capacitor.
The capacitance of a plate capacitor is calculated, ignoring fringe fields. But
these play the leading role in the pair of cables.

Fig. 4.21 Two small
square pieces of charged flat
electrodes and their fields. If
brought closely together, the
field strengths between them
add up. On the back, they
cancel

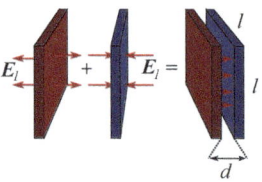

In contrast, the following transition is smooth and without jumps: Imagine small sections of two flat electrodes of a capacitor that are at a great distance from each other. In this case, as indicated in ■ Fig. 4.21, both have as many field lines on the front as on the back. For a surface charge density $Q_A = Q/A$, Gauss' theorem gives a field strength $E_1 = Q_A/(2\varepsilon)$ for an area element of the size $A = l^2$. The field between the two elements is the sum of the fields of electrodes:

$$E = 2\,E_1 = 2\frac{Q_A}{2\varepsilon} = \frac{Q}{\varepsilon \cdot A}\,.$$

If the electrodes are moved towards each other up to a distance d, the situation becomes as follows:
- Between the plates (as usual), the voltage is $U = (Qd)/(\varepsilon A)$.
- Beyond the plates, the fields cancel (also as usual): $E_{\text{outside}} = 0$.

The equality of the results of "extreme sphere" and "simply two plates." is a consequence of the following rule:

❯ Using Gauss' Law for Capacitors
When using Gauss' law, the capacitance of a set of electrodes must be calculated from all charges present. Apparent exceptions only occur if field components disappear in special geometries.

These exceptions are the spherical capacitor, its extreme case, the plate capacitor, and the coaxial capacitor (see ■ Problem 4.11). These form the vast majority of all technical applications. Therefore, the wrong impression can easily arise that the exceptions are the rule.

4.7 Problems

4.1 Find an expression for the potential of two point charges Q_1 and Q_2, located at points r_1 and r_2. Generalise the result to n point charges. Then, suggest corresponding expressions for the electric field starting from 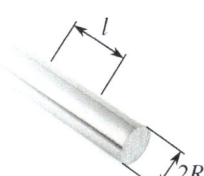 Eq. (3.1).

4.2 Why do electric field lines always approach a conductor at an angle of $90°$?

4.3 Please determine the progression of the potential energies $W_p(x)$ and $W_s(x)$ of the arrangements shown in ◘ Fig. 4.2 in an analytical manner.

4.4 Please determine the electric field inside and outside of the very long cylinder shown in ◘ Fig. 4.22 using Gauss' theorem. The cylinder has a radius R and a charge density ρ. It may be helpful to start with a piece of some length l (later, this will cancel). Is it easier to solve the problem using Poisson's and Laplace's equations?

4.5 In the late stage of large stars, carbon nuclei fuse. As ◘ Fig. 4.23 shows, this happens in the core. Two carbon nuclei can fuse to form a magnesium nucleus: $^{12}C + {}^{12}C \rightarrow {}^{24}Mg$. One can assume that atomic nuclei have a similar mass density and a homogeneous charge distribution and are roughly spherical. The radii can be estimated from the number of nucleons A (here 12 or 24) with $R \approx 1.07$ fm $\cdot A^{1/3}$. Also, $\varepsilon_r \approx 1$ may be assumed. What potential energy, expressed in electron volts, must be overcome for two nuclei to touch each other (The growth of internal potential energy in the course of the approach of the two carbon nuclei may be neglected.)? Please compare the potential energy of the charge carriers before and after fusion: By which factor does it increase?

◘ **Fig. 4.22** Illustrating Problem 4.4: a very long metal rod with radius R, from which a piece of length l is considered

Fig. 4.23 Illustrating
Problem 4.5: Cross section
of a red giant star. Carbon
fuses in the core, and helium
fuses in the shell around it

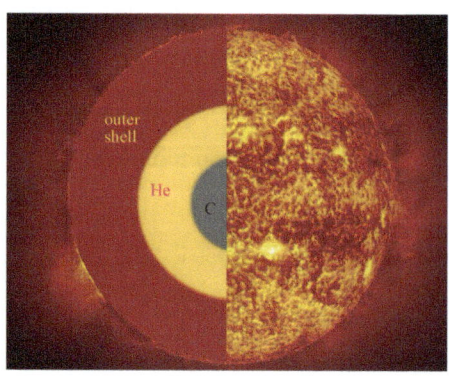

4.6 Please show that the capacitance of a plate capacitor is a limiting case of a spherical capacitor with spherical shells that are very close to each other if fringe effects can be neglected.

4.7 Please determine the area of a PET film capacitor ($\varepsilon_r = 3.3$) with a film thickness of $d = 1\ \mu m$ and a capacitance of $C = 10$ nF. Compare the area with the surface of a metal ball having the same capacitance.

4.8 Show that for non-trivial boundary conditions, the function $F(r, r')$ in Eq. (4.22) must satisfy the Laplace equation.

4.9 Show that the dipole moment of a system of point charges calculated according to Eq. (4.23) is independent of the choice of the coordinate system.

4.10 The value for the dipole moment of a water molecule, H_2O, is determined by measurement to be $p = 0.613 \cdot 10^{-29}$ Cm. Its two hydrogen atoms form an angle of $\phi = 104.45°$. The distance between the oxygen nucleus and the hydrogen nuclei is $a = 0.09584$ nm. Please calculate the dipole moment from these values provided that the two hydrogen atoms are completely ionised and compare the result with the measured value. What causes the difference?

4.11 For the inside of a conductor to be field-free, even if there are charges close by some of the conductor's charges must migrate to the surface. For a coaxial cable, this means that in the event of a large tension from the outside to the inside, there must be an excess of electrons at the surface of the inner conductor. Nevertheless, when calculating the conductor's resistance, the current density is assumed to be independent of the radius. How can this contradiction be resolved?

4.8 Solutions

4.1 To find the potential at a point not equal to $r_0 = (0, 0, 0)$, one has to move the coordinate system in the opposite direction. Then, the potential due to the charge Q_1 at r_1 is $\Phi = Q_1/(4\pi\varepsilon_0|r - r_1|)$. Since electric fields can be added, these are the correct formulas for many charges:

$$\Phi = \frac{Q_1}{4\pi\varepsilon_0|r - r_1|} + \frac{Q_2}{4\pi\varepsilon_0|r - r_2|} \quad | \text{ for 2 charges}$$

$$\Phi = \frac{1}{4\pi\varepsilon_0} \sum_{i=1}^{n} \frac{Q_i}{|r - r_i|} \quad | \text{ for } n \text{ charges,}$$

The same shift can be applied to the electric field:

$$E = \frac{Q_1(r - r_1)}{4\pi\varepsilon_0|r - r_1|^3} + \frac{Q_2(r - r_2)}{4\pi\varepsilon_0|r - r_2|^3} \quad | \text{ for 2 charges}$$

$$E = \frac{1}{4\pi\varepsilon_0} \sum_{i=1}^{n} \frac{Q_i(r - r_i)}{|r - r_i|^3} \quad | \text{ for } n \text{ charges.}$$

4.2 If a field line is perpendicular to the surface, the electric field has no component parallel to the surface. Otherwise, the charge carriers would move inside the conductor until this field component disappears. Since the charge carriers cannot leave the conductor, the argument does not apply to the vertical component of the electric field.

4.3 According to ◻ Eq. (4.3) for $Q_2 = Q_3 = Q/2$, the potential energy is given by

$$W = \frac{Q^2}{4\pi\varepsilon_0} \left(\frac{1}{2r_{12}} + \frac{1}{2r_{13}} + \frac{1}{4r_{23}} \right),$$

with $r_{23} = A$. If all charge carriers are aligned, then $r_{12} = |x - A/2|$ und $r_{23} = |x + A/2|$, and the potential energy is

$$W = \frac{Q^2}{4\pi\varepsilon_0}\left(\frac{1}{2|x - \frac{4}{2}|} + \frac{1}{2|x + \frac{4}{2}|} + \frac{1}{4A}\right).$$

Here, the amount dashes are crucial.

If the distance vector between the two more minor charges is perpendicular to the distance vector to the larger charge Q,

$$W = \frac{Q^2}{4\pi\varepsilon_0}\left(\frac{1}{\sqrt{x^2 + \frac{4^2}{4}}} + \frac{1}{4A}\right)$$

is the correct expression.

4.4 The cylindrical symmetry implies that the electric field can only have one radial component: $E = (E_r, E_\phi, E_z) = (E_r, 0, 0)$. Gauss' theorem can, therefore, be written in cylindrical coordinates for a radius $r \leq R$ and a line of length l in the following manner:

$$r\int_0^l\int_0^{2\pi} \varepsilon E_r \mathrm{d}\phi \mathrm{d}z = \int_0^r\int_0^l\int_0^{2\pi} \rho \mathrm{d}\phi \mathrm{d}z r \mathrm{d}r.$$

For a constant charge density, the result is a field strength that increases with the radius,

$$2\pi r l \varepsilon E_r = \pi l r^2 \rho \rightarrow E_r = \frac{\rho r}{2\varepsilon} \text{ (for } r < R),$$

where the increase due to $\varepsilon = \varepsilon_0\varepsilon_r$ decreases with polarizability (relevant for semiconductors). For radii $r > R$, the polarizability no longer affects the electric field strength, and the amount of charge remains constant. Outside of the conductor, the field strength decreases with the radius:

$$2\pi r l \varepsilon_0 E_r = \pi l R^2 \rho \rightarrow E_r = \frac{\rho R^2}{2r\varepsilon_0} \text{ (for } r > R).$$

If the material has a relative permittivity different from one, the field strength is non-continuous at the surface of the material.

The result for $r > R$ can generally not be formulated as the derivative of a potential function. Since $\int r^{-1}dr = \ln(r)$, there is no potential function, which is finite for $r \to \infty$. The use of the Laplace equation is, therefore, ineffective.

4.5 According to ◼ Eq. (4.13), the voltage that is needed to make the carrier of the elementary charge e touch one of the nuclei is

$$U = \frac{6e}{4\pi\varepsilon_0 \cdot 2R},$$

where $R = 1.07 \cdot A^{1/3}$ fm $= 2.45$ fm. The numbers add up to $U = 1.76$ MV. This value is equal to the energy in electron volts needed by the carrier of the elementary charge to touch the nucleus. A carbon nucleus needs six times the energy, i.e. $W = 10.6$ MeV.

Next, the internal potential energies shall be computed according to ◼ Eq. (4.16). For $\varepsilon_r = 1$ the potential within the nuclei is

$$\Phi_{inner}(r) = \frac{\rho}{2\varepsilon_0}\left(R^2 - \frac{r^2}{3}\right).$$

Using $dV = 4\pi r^2 dr$, the energy can be calculated to be

$$W_{pot} = \frac{1}{2}\int_0^R \rho \cdot \frac{\rho}{2\varepsilon_0}\left(R^2 - \frac{r^2}{3}\right) \cdot 4\pi r^2 dr = \frac{4\pi\rho^2 R^5}{15\varepsilon_0},$$

i.e. increasing with the fifth power of the radius. But the radius only increases with the cube root of the number of nucleons A. So the ratio of inner energies is

$$\frac{W_{pot}(^{12}C)}{W_{pot}(^{24}Mg)} = \left(\frac{R(^{12}C)}{R(^{24}Mg)}\right)^5 = \left(\frac{12}{24}\right)^{5/3} = 0.315.$$

There were two carbon nuclei before the merger. Both together have 63 % of the potential energy of the charges in the magnesium core. In other words, the reaction is electrically exothermic, nuclear forces must overcompensate the increase of internal energy.

4.6 Using $d = R_2 - R_1 \ll R_1$, ◘ Eq. (4.35) can be rewritten for small distances as

$$C = 4\pi\varepsilon_0 \frac{R_1 R_2}{R_2 - R_1} \approx \frac{4\pi\varepsilon_0 R^2}{d},$$

where $A = 4\pi R^2$ is the (almost identical) surface area of the spheres.

4.7 In this case,

$$a = \frac{Cd}{\varepsilon_0\varepsilon_r} = 3.42 \; 10^{-4} \; \text{m}^2 \; \text{(filmcapacitor)}$$

applies to the area a of the film capacitor. For the sphere, the radius is

$$R = \frac{C}{4\pi\varepsilon_0} = 98.9 \; \text{m (ball)},$$

which already indicates that the film capacitor makes better use of space. One can also calculate the ratio of the surfaces directly. The result

$$\frac{a_{\text{sphere}}}{a_{\text{filmcapacitor}}} = \frac{C\varepsilon_r}{4\pi\varepsilon_0 d} \approx 3 \cdot 10^8$$

confirms the impression.

4.8 From ◘ Eqs. (4.17, 4.21) is known that

$$\Delta\left(\frac{1}{4\pi|r - r'|}\right) = -\delta^3(r - r'),$$

Green's function is a solution for potentials that disappear at the boundary. If further terms are added, they must disappear if differentiated twice. So $\Delta F(r, r') = 0$ is required. This is the Laplace equation.

4.9 Shifting a coordinate system is equivalent to adding a fixed vector R_V. The dipole moment in the shifted reference system is therefore

$$P = \sum_i Q_i(r_i + R_V) = \left(\sum_i Q_i r_i\right) + R_V \sum_i Q_i,$$

which shows that the dipole moment in the new coordinate system is the same as the original one, provided that the system as a whole is neutral ($\sum_i Q_i = 0$).

4.10 The solution to ◘ Problem 4.9 proves the freedom to choose the coordinate system to determine the dipole moment. In this case, placing the oxygen atom in the coordinate origin and the hydrogen nuclei symmetrically around the x-axis is particularly advantageous. Then, the polarisation vector must be of the form $\boldsymbol{p} = (p, 0, 0)$. Then, the only component is $p = 2 \cdot e \cdot d \cdot cos(\phi/2)$. Here ϕ is the angle between the hydrogen atoms, and d is their distance from the oxygen nucleus. The value is $p = 1.88 \cdot 10^{-29}$ Cm. It is over three times as large as the experimentally determined value. Since $p = q \cdot d$ and d are known, Q must be less than e. The hydrogen atoms are only partially ionised. They kept a considerable proportion of the electrons to themselves.

4.11 There is undoubtedly an excess of electrons on the surface of the inner conductor. But this excess is so small under all conceivable conditions that it can be neglected. Here is a numerical example:

Assume a coaxial cable whose copper inner conductor has a radius from $r = 5$ cm and whose outer conductor starts at $R = 10$ cm. In between is an insulator with $\varepsilon_r = 2$ subjected to a voltage of $U = 100\,000$ V.

The charge per length Q/l on the surface due to the voltage U can be calculated as

$$U = \int_r^R E_{x'} \mathrm{d}x' = \int_r^R \frac{Q}{2\pi\varepsilon l x'} \mathrm{d}x' = \frac{Q}{2\pi\varepsilon l} \ln\left(\frac{R}{r}\right) \tag{4.36}$$

and with $Q = e$ leads to a number of

$$\frac{N_e}{l} = 1.00 \cdot 10^{14}\ \mathrm{m}^{-1},$$

electrons per meter on the surface.[8]

In comparison, the total number of conduction electrons, calculated from Avogadro's number N_A and the molar volume V_{mol} of copper[9] per length

---

8 The clever reader will notice that with ◘ Eq. (4.36) the formula for the capacitance of a coaxial cable has almost been found.

9 Per copper atom, there is one conduction electron.

$$\frac{N_e}{l} = \pi r^2 \frac{N_A}{V_{\mathrm{mol}}} \rightarrow \frac{N_e}{l} = 6.65 \cdot 10^{26}\ \mathrm{m}^{-1}$$

is so much larger that no significant effect on the conductivity can be expected.

References

1. Paul A. Tipler and Gene Mosca, Physik, 8. Auflage, Springer-Spectrum 2019, ISBN 978-3-662-58280-0
2. David J. Griffiths, Electrodynamics, 5. Edition Cambridge University Press 2023, ISBN 9781009397759
3. Juergen Schnakenberg, Elektrodynamik, Viley-VCH, Stuttgart 2009, ISBN: 978-3-527-40369-1
4. Siegfried Blume, Theorie elektromagnetischer Felder, 3. Auflage, Huethig, Heidelberg 1991, ISBN 978-3778520703
5. Kupfmueller, Theoretische Elektrotechnik, 20. Auflage, Springer Heidelberg 2017, ISBN 978-3-662-54837-0
6. Marco Leone, Theoretische Elektrotechnik, 2.Auflage, Springer Heidelberg 2021, ISBN 978-3658292072
7. David Dugdale, Essentials of Electromagnetism Springer New York 1993, ISBN 978-1-56396-253-0
8. J. B. Tatum, Dipole and Quadrupole Moments, University of California 2019 ▶ https://phys.libretexts.org/Bookshelves/Electricity_and_Magnetism/Electricity_and_Magnetism_(Tatum)
9. John David Jackson, Classical Electrodynamics, 3. Auflage, Wiley 1998, ISBN 978-0-471-30932-1

Second Special Case: Static Magnetic Fields

Contents

© The Author(s), under exclusive license to Springer-Verlag GmbH,
DE, part of Springer Nature 2024
M. Poppe, *Basic Electrodynamics in 6 Lessons*,
https://doi.org/10.1007/978-3-662-69143-4_5

Abstract Electrical machines use the properties of magnetic fields in ferromagnetic materials. This chapter describes the details of the energy transfer between the currents, the fields, and the electrons in the iron. In this manner, material properties are connected to machine efficiencies. It is shown that, depending on how magnetically active substances are distributed in space, the magnetic potential must fulfil different equations. This helps to understand why Boit-Savart's law for calculating magnetic fields is not always applicable. It is shown that torques act on closed current loops of any shape and that the dipole moments characterising the loops have similarities with electric dipole moments. Multipole expansions are introduced. The chapter also discusses the use and validity of magnetic circuits. It ends with some remarks on inductances.

Static electric fields do not affect magnetic fields

Magnetic fields are vortex fields like the velocity field of a cyclone shown in $\square$ Fig. 5.1. As long as no alternating electric fields are present, the field of the magnetic force, B ("the magnetic flux density") is exclusively caused by moving charges, resp. currents. Maxwell's equations reduce to $\nabla \cdot B = 0$ and $\nabla \times (\mu^{-1} B) = J$ for static fields. The static magnetic field is completely independent of the static electric field.

5.1 Energies in the Magnetic Field

The energy density of the magnetic field,

$$w_{\text{field}} = \frac{1}{2\mu_0} B^2 ,$$

(5.1)

$\square$ **Fig. 5.1** An Atlantic cyclone. Its field of air velocities is a vortex field. Therefore, the energy consumption for the Frankfurt - Greenland flight strongly depends on the route chosen. Kerosene can be saved by taking detours

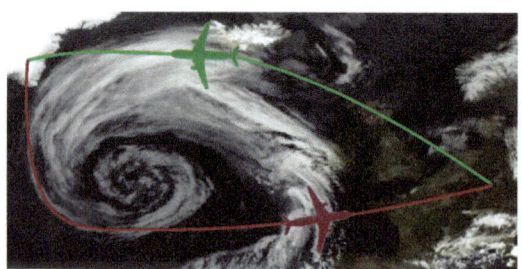

can be determined from the energy needed to build up the field of a current-carrying air coil. This relationship applies regardless of presence or absence any substances. It specifies the energy stored in the field (only).

In the presence of matter, ■ Eq. (5.1) can be modified to obtain the energy density

$$w_{\text{total}} = \frac{1}{2\mu_0\mu_r} B^2, \tag{5.2}$$

of the field and the substance it is exposed to together. Diamagnetic substances ($\mu_r < 1$) weaken the field by the generation of an opposing field fraction. Paramagnetic substances, on the other hand, can strengthen a magnetic field by releasing energy that is stored within the material. More than 99% of all known substances have a μ_r, which differs from one by less than 1%. Consequently, $w_{\text{field}} \approx w_{\text{total}}$ is the most frequent case. Very large deviations are found in ferromagnetic materials like iron, nickel, and cobalt, and their alloys and some of their compounds. Because of the outstanding technical importance, their influence on magnetic fields shall be examined in more detail:

In iron, small currents provoke a release of large energies

An iron core placed into a coil may increase the field **B** by factors up to two thousand, even though the current remains unchanged. Such materials also exhibit *hysteresis*. The strength of the field not only depends on the present current of the coil but also on the field's strength and direction of the past. This phenomenon has non-trivial consequences for the energy and power balance of systems with ferromagnetic substances, which shall be looked at next.

Assume a coil with a slow, linear increase in current from zero up to a point, where saturation[1] of the iron is reached. Then, the current is lowered back to zero. For the sake of simplicity, a constant core cross-section a, and a homogeneous field are assumed. Then, the magnetic flux is $\Phi_B = B \cdot a$, anolgous to that of a very long coil ($l \gg \sqrt{a}$). As shown in ■ Fig. 5.2, the magnetic field passes through the points $B(t = -t_0) = -B_R$, then $B(t = 0) = B_{\text{sat.}}$, and finally $B(t = t_0) = B_R$. The following three types of powers may now be distinguished.

---

[1] Saturation reached when the increase in the coil's field contribution is equal to that of the entire field.

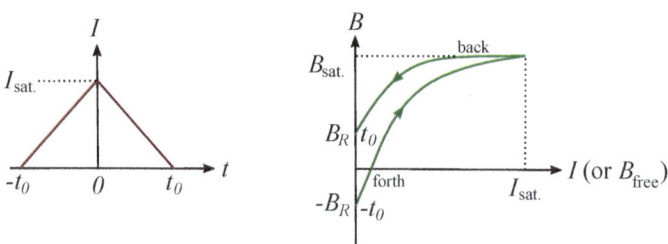

■ **Fig. 5.2** On the energy balance of magnetic field hysteresis in iron: An iron core is exposed to the field of a coil current I that increases and decreases linearly (shown left). The progression of the magnetic field strength is shown in the curve on the right. The horizontal axis may either be the current I or the field fraction B_{free}

$$P_I = U_{\text{ind.}} I = N \frac{d\Phi_B}{dt} I = NaI \frac{dB}{dt} \qquad \text{(power from the circuit)}$$

$$P_B = \frac{dW_{\text{field}}}{dt} = \frac{la}{\mu_0} B \frac{dB}{dt} \qquad \text{(power to the field) and} \qquad (5.3)$$

$$P_{Fe} = P_I - P_B = a\left(NI - \frac{l}{\mu_0} B\right) \frac{dB}{dt} \quad \text{(power to the iron core)}$$

These formulas simplify if written as a function of the field portion $B_{\text{free}} = B(I) = \mu_0 NI/l$ which is generated by the coil current alone. Then, all geometry factors can be forced to vanish by switching to power densities $p = P/V = P/(l\,a)$. The results

$$p_I = \frac{B_{\text{free}}}{\mu_0} \frac{dB}{dt} \qquad \text{(from the circuit)}$$

$$p_B = \frac{B}{\mu_0} \frac{dB}{dt} \qquad \text{(to the field) and} \qquad (5.4)$$

$$p_{Fe} = p_I - p_B = \frac{1}{\mu_0}(B_{\text{free}} - B) \frac{dB}{dt} \quad \text{(to the iron core)}$$

are much easier to interpret. In particular, the energy flow into the iron and back is clarified:

- At $t \approx -t_0$, near $B_{\text{forth}} \approx -B_R$, the iron absorbs energy from both the field and the circuit, from $B = 0$ only from the circuit.
- Starting at $B_{\text{forth}} = B_{\text{free}}$, the iron begins to release energy to the field.

- At $B_{\text{forth}} = B_{\text{back}} = B_{\text{sat.}}$, the iron has released the maximal amount of energy to the field.
- On the way back ($0 < t < t_0$) the iron, like the circuit, takes energy from the field.

An astonishing consequence arises for the internal energy of the iron:

⬤ Internal Energy of Iron

Magnetised iron has less internal energy than unmagnetised iron. In the absence of an external magnetic field, the minimum is at $B = B_R$. In the presence of an external field, this minimum is even undercut.

Now, it becomes clear why a compass needle, once magnetised, lasts for decades by retaining its magnetisation.

In 1881, Emil Warburg [1] determined the effect of hystereses on the efficiency of electrical machines and transformers. He found that the iron absorbs more energy while the magnetic field is built up than it releases during the field's decline. His findings are crucial for the efficiency of electrical machines. Here are the details:

The energy taken from the circuit while the magnetic field is being built up is

$$W_{\text{forth}} = \int_{-t_0}^{0} P_{\text{forth}}\mathrm{d}t' = \int_{-t_0}^{0} Na\frac{\mathrm{d}B_{\text{forth}}}{\mathrm{d}t'}I(t')\mathrm{d}t' = \int_{-t_0}^{0} Na\frac{\mathrm{d}B_{\text{forth}}}{\mathrm{d}t'}I_{\text{sat.}}\left(1 + \frac{t'}{t_0}\right)\mathrm{d}t'.$$

for the forward progression of the magnetisation. Partial integration gives

$$W_{\text{forth}} = NaB_{\text{sat.}}I_{\text{sat.}} - Na\frac{I_{\text{sat.}}}{t_0}\int_{-t_0}^{0} B_{\text{forth}}\mathrm{d}t'$$

and after a similar calculation

$$W_{\text{back}} = -NaB_{\text{sat.}}I_{\text{sat.}} + Na\frac{I_{\text{sat.}}}{t_0}\int_{0}^{t_0} B_{\text{back}}\mathrm{d}t'$$

for the way back. Adding the energies leaves the integral terms. These can be transformed into integrals over the current.[2] The sum may then be written as

$$W = W_{\text{forth}} + W_{\text{back}} = Na \int_0^{I_{\text{sat.}}} (B_{\text{back}} - B_{\text{forth}})dI' .$$

The time t_0 no longer appears in this formula. So, it is irrelevant how quickly everything happens, even details of the progression of the current over time do not matter. The neatest form of the result is obtained if, instead of the current, the field component B_{free} is taken as the integration variable. The result of the variable transformation

$$w = \frac{W}{al} = \frac{1}{\mu_0} \int_0^{B_{\text{free,sat.}}} (B_{\text{back}} - B_{\text{forth}})dB'_{\text{free}} \tag{5.5}$$

shows that if the variables are chosen wisely, the energy loss density w can be read off the graph $B(B_{\text{free}})$. It is the area between the outward and return path, divided by μ_0 (■ Fig. 5.3). The energy loss leads to iron heating as $w = w_{Fe}$ applies here.

So far, only the part of the hysteresis curve where I is positive has been considered. Of course, the formulas also apply for $I \leq 0$. The result for the overall curve, which represents a complete remagnetization cycle, is shown in ■ Fig. 5.3. The area enclosed by the hysteresis curve is, up to the factor μ_0, just equal to the loss of energy density per magnetisation cycle.

■ **Fig. 5.3** The graph for a complete magnetic reversal cycle. The energy loss density w_{Fe} within an iron core is determined by the enclosed area, divided by μ_0

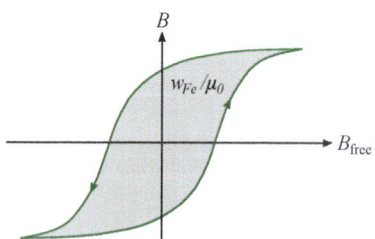

---

2   Using $I'_{\text{forth}} = I_{\text{sat.}}[1 + (t'/t_0)]$ for the path forward and a similar formula for the return path.

Magnets from ferromagnetic alloys with large enclosed areas are called *hard magnets.* They are particularly suitable for permanent magnets and compass needles. For the construction of transformers and those parts of electrical machines in which remagnetisation takes place, one will use *soft magnets*, i.e., prefer those with a narrow hysteresis curve.

> **Hard and Soft Iron**
> The best alloy for compass needles is the worst for electrical machines, and vice versa.

5.2 Vector Potential and the Poisson Equation

If the shape of a magnetic field is known a priori, the Ampere-Maxwell (3.22) law suffices to calculate the field B (the "magnetic flux density"). In all other cases, it is necessary to take a formal detour, which is analogous to using the potential in electrostatics. The detour begins at the *magnetic vector potential* A and ends with the so-called *Biot-Savart law*, which (under certain boundary conditions) allows the magnetic field to be calculated via integration at any point in space.

The magnetic potential can be gauged

The magnetic vector potential A is defined by

$$B = \nabla \times A .\tag{5.6}$$

The name "potential" refers to the analogy to electrostatics. It is not derived from the term potential energy.

For a given magnetic field, there are several vector potentials because portions of A that vanish under the operation $\nabla\times$ do not contribute to B. Because of $\nabla \times (\nabla\Psi) = 0$ one may conclude: If A can be used to determine the correct field of the magnetic force B, then for an arbitrary function Ψ,

$$A_\Psi = A + \nabla\Psi\tag{5.7}$$

must lead to exactly the same magnetic field. Choosing a specific $\nabla\Psi$ will *gauge* the potential, $\nabla\Psi$ is called *gauge field*.

Dynamic theories whose potentials can have gauge fields are called *gauge theories*. They play an outstanding role in the search for a common formulation of all forces. Gauge field theories describe electromagnetism, the so-called weak interaction, and the nuclear force. In magnetostatics, the *Coulomb gauge*

$$\nabla \cdot A = 0 \quad \text{(Coulomb gauge)} \tag{5.8}$$

is the most popular one. Therefore, in this chapter, the vector potential is also defined as a pure vortex field.

Matter defines the choice of the differential equation

Just like a charge distribution $\rho(r')$ defines a potential $\Phi(r')$ that gives a field $E(r')$, a current density distribution $J(r')$ defines a magnetic vector potential $A(r')$ that gives a field $B(r')$.

Here, Ampere's law (3.25) paves the way as follows:

$$
\begin{aligned}
B &= \nabla \times A & &| \, \mu^{-1} \cdot \\
\mu^{-1} \cdot B &= \mu^{-1} \cdot (\nabla \times A) & &| \, \nabla \times \\
\nabla \times (\mu^{-1} \cdot B) &= \nabla \times \left[\mu^{-1} \cdot (\nabla \times A) \right]
\end{aligned}
$$

So with the help of ◼ Eq. (3.12) one obtains the *general potential equation for magnetic fields in matter*

$$J_{\text{free}} = \nabla \times \left[\mu^{-1} \cdot (\nabla \times A) \right] \quad \text{(always valid in the static case)}, \tag{5.9}$$

which, however, is generally very difficult to handle. Therefore, the following special cases are worth looking into. Using $\nabla \times (\nabla \times A) = \nabla(\nabla \cdot A) - \Delta A$ they may be calculated as follows:

If the substance does not alter the direction of the magnetic field,

$$J_{\text{free}} = -\mu^{-1} \Delta A + (\nabla \mu^{-1}) \cdot (\nabla \times A) \quad \text{(if } \mu \text{ is a scalar)}$$

remains. If μ is a constant number, then the potential equation simplifies to the *Poisson equation for magnetic fields*

$$\mu J_{\text{free}} = -\Delta A \quad \text{(if } \mu \text{ is a constant number everywhere)} . \tag{5.10}$$

The formal similarity to that of the electric field allows some knowledge gained in electrostatics to be used in magnetostatics. In the absence of currents, the *Laplace equation for magnetic fields*

$$\Delta A = (0, 0, 0) \quad \text{(if } \mu \text{ is a constant number everywhere)} \qquad (5.11)$$

is the right choice. Both vector equations stand for three individual equations.

5.3 The Biot-Savart Law

With the help of this law, one can calculate, the magnetic field close to a current-carrying conductor. Mathematically, it is a consequence of a particular solution of the Poisson equation. It only applies in the absence of field-rotating materials or material inhomogeneities. Here is the path to the law:

If A disappears at infinity, the solution of the Poisson equation is known from ◘ Eq. (4.15):

$$A = \frac{\mu}{4\pi} \int_V \frac{J(r')}{|r - r'|} dV' \qquad (5.12)$$

According to the Heine-Cantor theorem and the chain rule one gets for a resting volume V

$$B = \nabla \times A = \frac{\mu}{4\pi} \int_V \nabla \times \left(\frac{J(r')}{|r - r'|} \right) dV'$$

$$= \frac{\mu}{4\pi} \int_V \left[\frac{1}{|r - r'|} \nabla \times J(r') - J(r') \times \nabla \left(\frac{1}{|r - r'|} \right) \right] dV'.$$

The term $\nabla \times J(r')$ is equal to zero because the nabla operator ∇ only acts on r, not on the integration variable r' (compare $\partial f(x')/\partial x = 0$). This leaves the gradient of $1/|r - r'|$ to be calculated to get Biot-Savart's law

$$B(r) = \frac{\mu}{4\pi} \int_V \frac{J(r') \times (r - r')}{|r - r'|^3} dV'. \qquad (5.13)$$

If the current density is limited to form a thin line carrying a current I, the volume integral can, as shown in the solution to ◘ Problem 5.3, be reduced to a path integral along that line. It reads for this case

■ **Fig. 5.4** Variables used in Biot-Savart's law: The field B ("magnetic flux density") at location P results from integration over elements ds' of the current carrying line

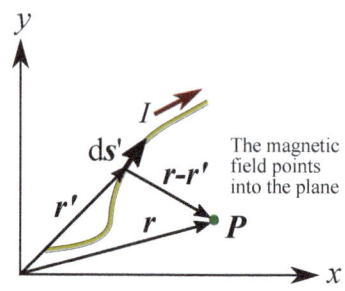

The magnetic field points into the plane

5

$$B(r) = \frac{\mu}{4\pi} I \int_s \frac{ds' \times (r - r')}{|r - r'|^3} \,.$$ (5.14)

The variables are sketched in ■ Fig. 5.4.

Often, it is convenient to determine the vector potential prior to the field. For a thin wire, this is a special case of ■ Eq. (5.12):

$$A = \frac{\mu}{4\pi} I \int_s \frac{ds'}{|r - r'|} \,.$$ (5.15)

Finally, note that Biot-Savart's law may only be used if there are no field-rotating substances nearby, and if the entire environment has the same μ_r.

Example: The Field of an Arc Segment

■ Figure 5.5 shows one of the few analytically calculable examples for the application of Biot-Savart's law. The field in the centre of the quarter arc with the radius of curvature R is to be calculated for a current I following the path shown in ■ Fig. 5.5. The integration over the arc segments ds' can be reduced to the integration over the angle ϕ'. Then, the following identities may be used

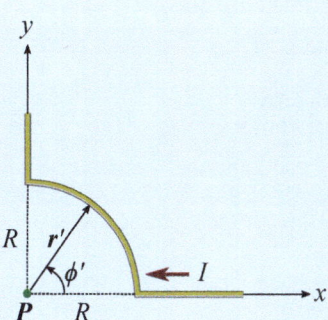

Fig. 5.5 A wire carrying a current I is bent to form a quarter arc. Here, the magnetic field at point P is to be calculated

$$r = (0, 0, 0)$$
$$r' = R(\cos\phi', \sin\phi', 0)$$
$$r - r' = -R(\cos\phi', \sin\phi', 0)$$
$$\mathrm{d}s' = \frac{\partial r'}{\partial \phi'}\mathrm{d}\phi' = R(-\sin\phi', \cos\phi', 0)\mathrm{d}\phi'$$

to write down the law for this application. The result

$$B = \frac{\mu}{4\pi}I\frac{1}{R}\int_{0}^{\pi/2} \left[(-\sin\phi', \cos\phi', 0)\mathrm{d}\phi'\right] \times (-\cos\phi', -\sin\phi', 0)$$
$$= \left(0, 0, \frac{\mu I}{8R}\right)$$

is a field in the z direction, i.e. pointing out from the page (see also □ Problem 5.6). The straight parts of the path give no contribution to the magnetic field at the point $P = (0, 0, 0)$ because for them $\mathrm{d}s \parallel r'$ applies.

In most cases, magnetic fields have to be calculated numerically

For a limited number of geometries, analytical solutions of □ Eq. (5.14) are known. These are shown in □ Fig. 5.6 and summarised in □ Eq. (5.16).

5

◻ **Fig. 5.6** Some applications
of Biot-Savart's law. From top
to bottom: current arc, ring
coil, half ring, half square,
rectangle. The results for the
field **B** at the locations labelled
P are shown in ◻ Eq. (5.16)

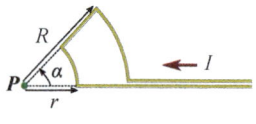

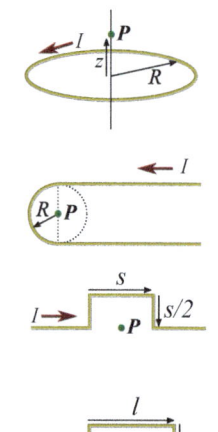

Examples for Magnetic Fields (see Fig. 5.6)

Type	Location	Field	Direction
current arc	circle center	$B = \dfrac{\mu I \alpha}{4\pi}\left(\dfrac{1}{r} - \dfrac{1}{R}\right)$	into the sheet
ring coil	central axis	$B = \dfrac{\mu I}{2}\dfrac{R^2}{\left(R^2 + z^2\right)^{(3/2)}}$	z – axis
half ring	circle center	$B = \dfrac{\mu I}{2\pi R}$	out of the sheet
half square	center of square	$B = \dfrac{\sqrt{2}\mu I}{\pi s}$	into the sheet
rectangle	center	$B = \dfrac{2\mu I}{\pi}\sqrt{\dfrac{1}{l^2} + \dfrac{1}{k^2}}$	into the sheet

$$(5.16)$$

Even small variations of the geometries can considerably increase the computational effort. For example, the calculation of the magnetic field of a circular coil at a location that is not on the central axis leads to an elliptic integral [2].

Analytical solutions to the Poisson equation are known for a limited number of geometries. A comprehensive compilation can be found in [3]. The vast majority of technical applications, however, require numerical methods. Descriptions of these methods can be found in [4, 5].

Vector Potential and Magnetic Field of a Long Wire

The vector potential A and the field B at a distance r from a very long, straight, thin cable shall be determined (see ◻ Fig. 5.7). For the vector potential of an infinitely long wire, the solution of ◻ Eq. (5.15) diverges. Therefore, one will first consider a line of finite length $2L$. Afterwards, it is examined what happens when L becomes very large.

The vector potential $A = (0, 0, A_z)$ is given by

$$A_z = 2 \cdot \frac{\mu}{4\pi} I \int_0^L \frac{dz'}{\sqrt{r^2 + z'^2}} .$$

The antiderivative for the above integral is known to be $\ln\left(\sqrt{z'^2 + r^2} + z'\right)$ for positive r and z' (it is more complicated for negative values of z'). Therefore,

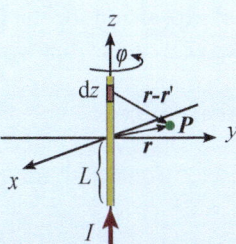

◻ **Fig. 5.7** Variables used to calculate the vector potential and the magnetic field of a long wire at the point P in the xy plane at a distance r. The current is assumed to flow in the direction of the z-axis. The field is made up of contributions from the line elements dz

$$A_z = 2 \cdot \frac{\mu}{4\pi} I \ln\left[\left(\frac{L}{r}\right) \cdot \left(\sqrt{1 + \frac{r^2}{L^2}} + 1\right)\right] \quad \text{(always)}$$

$$A_z = 2 \cdot \frac{\mu}{4\pi} I \ln\left(\frac{2L}{r}\right) \qquad\qquad \text{(for } L \gg r\text{)},$$

emerge for the vector potential. A diverges for large L. Nevertheless, it leads to the correct result for B: The only non-vanishing component of $\nabla \times (0, 0, A_z(r))$ is the vector component encircling the wire

$$B_\varphi = -\frac{\partial A_z}{\partial r} = -\frac{\mu I}{2\pi} \frac{\partial \ln(2L/r)}{\partial r} = \frac{\mu I}{2\pi r},$$

which no longer depends on L. It approaches zero for $r \to \infty$. A comparison with the integral form of Ampere's law (3.25) shows that the result is correct.

5.4 Forces in the Magnetic Field

All forces in magnetic fields can be traced back to the Lorentz force. Consequently, this section is about moving charges, resp. currents traversing magnetic fields.

Closed current loops are not displaced by homogeneous magnetic fields

Starting from a general expression for the Lorentz force on an arbitrary current distribution, the special cases "thin wire" and "closed current loop" shall be derived.

In general, the Lorentz force is

$$F = Qv \times B \qquad \text{(one charge)}$$

$$F = \sum_{i=1}^{N} Q_i v_i \times B \quad \text{(many discrete charges)}$$

$$F = \int_V \rho v \times B \, dV' \quad \text{(continuously distributed charge density)}$$

$$\to F = \int_V J \times B \, dV' \quad \text{(continuously distributed current density)}$$

If the current density J is limited to a thin wire, the integral over the volume V' is reduced to a path integral. The force acting between the points a and b of such wire is

$$F = I \int_a^b (\mathrm{d}s \times B) \quad \text{(thin wire)} \tag{5.17}$$

where the current I is the product of the current density and the cross-section of the wire (see ◙ Problem 5.3). If the magnetic field is homogeneous and has the same direction everywhere, it can be taken out of the integral.

An interesting special case arises if the wire is closed to form a current loop because the integral over a closed path is zero. One may deduce that according to

$$F = I \cdot \left(\oint \mathrm{d}s \right) \times B = I \cdot (0) \times B = 0 \,,$$

the net force on a closed loop current is zero in a homogeneous magnetic field. The loop stays where it is. Nevertheless, the magnetic field is not ineffective, as shown next:

Ring currents generate magnetic dipoles

Currents in closed conductor loops tend to to align themselves with a magnetic field in such a way that the charge carriers only move in a plane perpendicular to this field. Next, the associated torque τ will be determined from the Lorentz force $F = Qv \times B$. ◙ Figure 5.8 shows a thin wire carrying a current I, and placed in a constant magnetic field in the z-direction ($B = (0, 0, B)$). The line is located in a plane inclined by an angle θ around the x-axis. The torque by a carrier of the charge Q at the directed distance $R' = (0, R_y, R_z)$ from the z-axis is $\tau = R' \times F = R \times (Qv \times B)$. Here, v is the charge carrier velocity. Knowing the torque due to a single charge helps to find expressions for charge distributions, charge densities and currents.

For a continuous current distribution J, the torque must be calculated as a volume integral. The general formula is

$$\tau = \int_V R' \times J \times B \, \mathrm{d}V' \,. \tag{5.18}$$

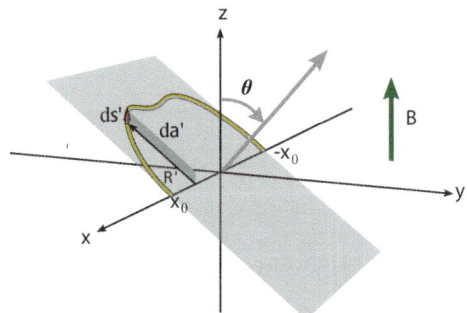

Fig. 5.8 A current-carrying thin wire in a plane which is inclined at an angle θ about the x-axis. Each piece ds at a distance R to the x-axis contributes to the torque

If the magnetic field has the same strength and direction everywhere, B can be removed from the integral. Also, if all charge carriers are located within a thin wire, the volume integral can be replaced by a path integral (see ■ Problem 5.2):

$$\boldsymbol{\tau} = I \left(\int_{-x0}^{x_0} \boldsymbol{R}' \times \mathrm{d}\boldsymbol{s}' \right) \times \boldsymbol{B}$$

The vector product $\boldsymbol{R}' \times \mathrm{d}\boldsymbol{s}'$ gives a vector d$\boldsymbol{a}'$, which is perpendicular to the plane of the wire, and its amount is equal to of the parallelogram spanned by $\boldsymbol{R}'$ and d$\boldsymbol{s}'$ (see ■ Fig. 5.8). The integral over the surface elements d$\boldsymbol{a}'$ results in the surface vector for the area $\boldsymbol{a}$ spanned by the wire and the x-axis:

$$\int_{-x0}^{x_0} \boldsymbol{R}' \times \mathrm{d}\boldsymbol{s}' = \int_{-x0}^{x_0} \mathrm{d}\boldsymbol{a}' = \boldsymbol{a} \, .$$

Consequently,

$$\boldsymbol{\tau} = I \boldsymbol{a} \times \boldsymbol{B} \tag{5.19}$$

The torque determined in this way depends on the choice of axis: the further away it is from the piece of wire, the larger $\boldsymbol{a}$ becomes.

However, if the wire is completed to form a closed loop, the torque no longer depends on the axis's choice. If one completes the loop for the wire shown in ■ Fig. 5.8 by a piece along the x-axis, the torque remains unchanged (along this axis, the distance to this $R = (0, 0, 0)$). The independence of the choice of axis is the result of a cancellation. Whatever contribution will be added to the torque from one piece of the wire will be compensated by subtraction from another one.

For a closed conductor loop, the torque given by ■ Eq. (5.19) is independent of the reference system. Therefore, it makes sense to think of it as an interaction of the magnetic field based on the loop's property called *magnetic dipole moment*, **m**:

$$\tau = m \times B = Ia \times B . \tag{5.20}$$

Since both τ and B are directly measurable quantities, the first equal sign in ■ Eq. (5.20) is used as the definition of the dipole moment **m** and the second is used for its calculation.

Finally, the role of the axis of rotation should be clarified. To illustrate the point, the components of torque for the example outlined above (including the loop completion along the x-axis) are written down explicitly:

$$\tau = BI \int_{-x_0}^{x_0} (R_z ds_x, R_z ds_y - R_y ds_z, 0)$$

The first component $\tau_x = R_z ds_x$ is the torque about the x-axis. However, as long as $\tau_y \neq 0$, a larger torque can be achieved with a free choice of axes. Consequently, if the loop can move freely, it will follow the largest torque possible. I will be rotated around an axis that points parallel to the torque vector and which divides the conductor loop into two halves.

Dipole moments can be added

The origin of the torques acting on conductor loops is the Lorentz force. This reverses its direction when the current direction is reversed. ■ Figure 5.9 shows the division of a current carrying ring into two halves. The Lorentz force of the new straight pieces of conductor exactly balance: they add up to zero. This would happen to any other line of division. Consequently, a current loop can be divided into many loops without any effect on the dipole moment. This indicates that one can construct larger dipoles by simply adding small individual dipoles.

The possibility of adding dipole moments arbitrarily can also be justified formally: The dipole moment **m** is a vector. It is the product of two quantities,

■ **Fig. 5.9** For the combination of dipole moments: a current-carrying ring is replaced by two half rings

I and a, which can both be added up (linearly). Therefore, dipole moments can also be added. This can be done even if they are tilted towards each other. Hence, one can obtain the dipole moment of an arbitrarily shaped current loop by dividing it into small elements and adding up the individual dipoles.

Microscopic dipoles create a macroscopic field

Solids can generate a magnetic field. Permanent ring currents inside their atoms and molecules create magnetic dipole moments. Therefore, almost every atom has a small resulting dipole moment m_1 of its own. The relationship between atomic dipole moments and the overall field shall be investigated next. As a "collateral benefit," it will become clear why a quantity appearing in Maxwell's equations is sometimes called *magnetic dipole density*.

■ Figure 5.10 shows a macroscopic cube of side length l containing a large number N of microscopic ring currents I_1. The resulting magnetic field can be traced back to the internal dipoles using the following considerations:

- Area vector, dipole moment and magnetic field are in parallel.
- Each dipole has a dipole moment $m_1 = I_1 a_1$.
- The area of the dipole is $a_1 = l^2/N^{2/3}$
- Long coils with $N^{1/3}$ turns are created. The magnetic field of each coil is $B_{\text{bound}} = \mu_0 I_1 N^{1/3}/l$. The index "bound" indicates that all currents are bound in the material.

The microscopic currents and surfaces may now be eliminated. As a result, the overall magnetic field emerges as being proportional to the dipole density $m_1 N/V$,

$$\boldsymbol{B}_{\text{geb.}} = \mu_0 \frac{N}{l^3}\boldsymbol{m}_1 = \mu_0 \frac{N}{V}\boldsymbol{m}_1, \tag{5.21}$$

which is traditionally denoted by $\boldsymbol{M} = \boldsymbol{B}_{\text{bound}}/\mu_0$. Based on these considerations $\boldsymbol{M}$ is often called *magnetic dipole density*.[3]

---

3   In this book, the term "magnetization" is reserved for the process of aligning the dipoles. In this way, the process is distinct from its result.

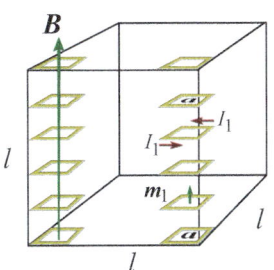

■ Fig. 5.10 For the derivation of ■ Eq. (5.21): A macroscopic cube with side length l, the material of which creates a magnetic field **B**, which is generated by many dipoles m_1. The latter arise from ring currents I_1 around areas of size a_1

❗ Model Limitations

The term dipole density should not be taken too literally because the actual number of dipoles bound in the atoms is a large multiple of the N. For almost every internal dipole, there is a second dipole of the same size but in the opposite direction. What is called m_1 here is the vector sum of many subatomic individual components. For example, in the case of iron, these are the dipole of the nucleus, the 26 individual dipole moments of the electrons, the 12 dipoles, which are due to the orbital movement of the inner shell electrons come as well as a further number of up to 8 dipoles of the outer electrons, whose orientation depends on the the the chemical bonding of the iron.

Beyond all simplified models of solids, if one thinks of B_{bound} being the portion of the magnetic field caused by bound charges, one is always correct.

5.5 Magnetic Multipoles

Magnetic fields and their effect on currents can be calculated analytically only in a few cases. One of the most popular procedures for the approximate calculation of a magnetic field is the *multipole expansion*.

The starting point is, as for the electric field, the series expansion of the vector potential. The variables used in this case are shown in ■ Fig. 5.11. Then, using ■ Eq. (5.15) the vector potential is written as

$$A = \frac{\mu}{4\pi} I \oint \frac{ds'}{\sqrt{(R - r')^2}} = A_0 + A_1 + \cdots \tag{5.22}$$

with

■ Fig. 5.11 Variables used to calculate the magnetic potential at point P at a distance R from the center of a closed current loop

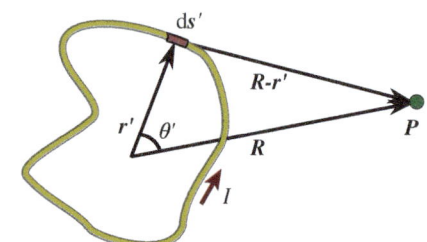

$$A_n = \frac{\mu}{4\pi R} \frac{I}{R^n} \oint (r')^n P_n(\cos\theta')ds' .$$

It's worth taking a closer look at the first terms:

$$A_0 = \frac{\mu}{4\pi R} I \oint ds' \qquad \text{(monopole term)}$$

$$A_1 = \frac{\mu}{4\pi R^2} I \oint r'\cos\theta' ds' \qquad \text{(dipole term)} \qquad (5.23)$$

$$A_2 = \frac{\mu}{4\pi R^3} I \oint \frac{r'^2}{2}\left(3\cos^2\theta' - 1\right) ds' \quad \text{(quadrupole term)}$$

The monopole term disappears since the path integral along a closed loop is always zero.

The following strategy will be followed to determine the dipole term: It is known that the torque of a plane current loop is proportional to its enclosed area. Since the dipole term contains a ring integral, it may be a good idea to use Stokes' theorem to rewrite the path integral as a surface integral. Then, one can check whether the area spanned by the current loop can be determined in this manner.

First, the dipole term used in ■ Eq. (5.23) is rewritten as a function of the vectors involved: $r'\cos\theta' = r' \cdot e_R$. The formula for the potential now looks like this:

$$A_1 = \frac{\mu}{4\pi R^2} I \oint_{\partial a} r' \cdot e_R ds'$$

Then, a formula derived from Stoke's theorem (see ■ Problem 5.2) can be used to obtain

$$\int_a da' \times (\nabla f) = \oint_{\partial a} f \, ds'$$

where $f = r' \cdot e_R$. Because of $\nabla(r' \cdot e_R) = e_R$[4] the surface integral is rewritten as

$$\int_a da' \times [\nabla(r' \cdot e_R)] = \int_a da' \times e_R = a \times e_R = \oint_{\partial a} (r' \cdot e_R) ds'.$$

This shows that the dipole term of the multipole expansion

$$A_1 = \frac{\mu}{4\pi R^2} I \, a \times e_R \tag{5.24}$$

is proportional to the area a enclosed by the current.

Furthermore, it becomes obvious that the current loop does not need to be in one plane. At no stage, the derivation of ◻ Eq. (5.24) needed this assumption. This confirms the finding that the magnetic dipole moment, even an arbitrarily shaped conductor loop can be calculated as $m = I a$. The dipole term of the potential expansion now reads

$$A_1 = \frac{\mu}{4\pi R^2} m \times e_R = \frac{\mu}{4\pi R^3} m \times R. \tag{5.25}$$

The dipole field is finally given by $B = \nabla \times A$. The gradient can be calculated as follows[5]

$$\nabla \times \frac{1}{R^3}(m \times R) = \left(\nabla \frac{1}{R^3}\right) \times (m \times R) + \frac{1}{R^3}\nabla \times (m \times R)$$

$$= -\frac{3}{R^4} e_R \times (m \times R) + \frac{2}{R^3} m$$

$$= \frac{1}{R^3}[-3e_R \times (m \times e_R) + 2m]$$

$$= \frac{1}{R^3}[3e_R(e_R \cdot m) - m]$$

---

4 e_R starts from the fixed point where A_1 is calculated, to the fixed point around which r' is calculated. So, the components of this vector are constant.

5 Here, $\nabla \times (m \times R = m(\nabla \cdot R) - (m \cdot \nabla)R = 3m - m$ is used.

and leads to

$$\boldsymbol{B} = \nabla \times \boldsymbol{A} = \frac{\mu}{4\pi R^3} \left[3(\boldsymbol{m} \cdot \boldsymbol{e}_R)\boldsymbol{e}_R - \boldsymbol{m} \right] . \tag{5.26}$$

A comparison with ◘ Eq. (4.29) shows that the magnetic field of the magnetic dipole has the same shape as the electric field of the electric dipole.

5.6 Inductances

The inductance L is the ratio of the magnetic flux Φ_B to its generating current:

$$L = \frac{\Phi_B}{I} \tag{5.27}$$

According to the Faraday-Henry law, a change in flux causes the induction of a voltage into a conducting loop. The inductance also determines which voltage is caused by a current change induced into a conductor loop.

A large inductance is always desired when a large amount of energy is to be stored in a magnetic field. A small inductance helps if the current shall flow independently of magnetic fields.

Coils use each conductor loop twice

If N loops made from a single wire are stuck on top of each other, the magnetic flux increases by a factor of N. Also, according to the Faraday-Henry law, the voltage induced into the series of loops is N times as large as that in a single loop. Consequently, N loops in series have N^2 times the voltage of a single loop. The well-known formula

$$L = \frac{\mu a}{l} \cdot N^2 \quad \text{(long coil)} \tag{5.28}$$

for the inductance of a long coil contains the factor N^2. And for this very reason, components that rely on the effect of magnetic fields usually have windings with many conductor loops placed one on top of the other. The name "coil" has become a synonym for components using magnetic fields.

Inductances of cables are determined by material and geometry

The magnetic flux Φ_B of a very thin, very long pair of wires can can be determined from the area between the wires. It is assumed that the radii r of the wires shown in ◘ Fig. 5.12 are much smaller than their distance d from each other. The flux is obtained as the sum of the flux fractions from the two wires

$$\Phi_B = 2 \cdot \int_r^{d-r} \frac{\mu I l}{2\pi y'} dy' = \frac{l\mu I}{\pi} \cdot \ln\left(\frac{d-r}{r}\right) \quad \text{(pair of thin wires)} \quad (5.29)$$

where $\mu = \mu_0 \mu_r$ refers to the material between the wires. If a small flux and, therefore, a small inductance is desired, then using wires with a large diameter is the first choice. If the inductance is bound to precisely have a given value because of the logarithmic singularity at $r \to 0$, the wires should also not be too thin.

If the tolerance of the value for the inductance is to be small, metal strips of height h can be used instead of round wires. Then, the singularity disappears, as shown in ◘ Problem 5.10. The strips can be as thin as desired. The magnetic flux of the arrangement shown in ◘ Fig. 5.12 is then according to the solution to ◘ Problem 5.10

$$\Phi_B = 2 \cdot \int_0^d \frac{l\mu I}{\pi h} \arctan\left(\frac{h}{2y'}\right) dy'$$

$$= \frac{l\mu I}{\pi} \left[\ln\left(\frac{\sqrt{h^2 + 4d^2}}{h}\right) + \frac{2d}{h} \cdot \arctan\left(\frac{h}{2d}\right) \right] \quad \text{(metal strips)}.$$

$$(5.30)$$

Its progression as a function of the normalised distance d/h is given by the term in the square brackets in ◘ Eq. (5.30). ◘ Figure 5.13, shows the inductance per length ($L/l = \Phi_B/(l \cdot I)$) as a function of the normalised distance y/h.

◘ **Fig. 5.12** A double line consisting of narrow metal strips (left), and a double line made of thin wires (right)

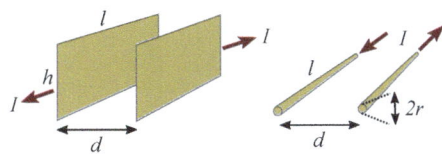

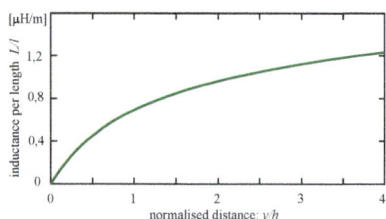

Fig. 5.13 Inductance (L/l) of two narrow bands of height h as a function of distance y of the bands. Here, $\mu = \mu_0$ is assumed

5.7 Magnetic Circuits

Magnetic circuits stand for an algorithm to determine the strengths of magnetic fields in ferromagnetic constructions. Its most popular applications are transformers and electrical machines. Magnetic circuits map magnetic fields onto electric circuits. In this manner, algorithms for calculating currents can be used to determine the strengths of magnetic fields.

The basis of using magnetic circuits is formed by analogies between currents, electric fields and magnetic fields. Only one of the equations used ($\nabla \cdot \boldsymbol{B} = 0$) is universally valid. Consequently, the use of magnetic circuits are restricted to static or quasi-static problems. The analogies are most easily visible in traditional notation:

law for magnetic fields		formally similar law	
$\nabla \cdot \boldsymbol{B} = 0$ (always)	$\leftrightarrow$	$\nabla \cdot \boldsymbol{J} = 0$	(statics)
$\oint \boldsymbol{H} \cdot \mathrm{d}s' = I$ (statics)	$\leftrightarrow$	$\int \boldsymbol{E} \cdot \mathrm{d}s' = -U$	(statics)

Some technical terms used to this day refer to these similarities. The "magnetic flux density" $\boldsymbol{B}$ addresses the current density $\boldsymbol{J}$, and the term "magnetic field" $\boldsymbol{H}$ is based on the analogy between Ampere's law and the definition of electrical voltage in an electrical circuit with a point-like voltage source: $U = -\oint \boldsymbol{E}\mathrm{d}s$.

Furthermore, there is an analogy between ohmic resistance and properties of magnetic material: An analysis of the refraction laws for magnetic fields (3.27) shows that magnetic field lines are almost exactly parallel to the surfaces of ferromagnetic materials (see ■ Problem 5.8). Consequently, the geometry of the material determines the shape of the field lines. And the magnetic flux $\Phi_B = \int_a \boldsymbol{B} \cdot \mathrm{d}a'$ within the materials is almost constant. For a magnetic field passing through several discrete parts and following a path s along the field lines, one gets a formula

$$NI = \oint_s \frac{\boldsymbol{B}}{\mu_0 \mu_r} \mathrm{d}\boldsymbol{s}' \approx \sum_i \frac{B_i s_i}{\mu_0 \mu_{ri}} = \sum_i \frac{B_i s_i a_i}{\mu_0 \mu_{ri} a_i} = \Phi_B \sum_i \frac{s_i}{\mu_0 \mu_{ri} a_i},$$

that resembles Ohm's law. $U = (\sum_i R_i)I$. It is, therefore, written in the form

$$U_M = \Phi_B \sum_i R_{M,i} \qquad (5.31)$$

and referred to as *Hopkinson's law*. Here, $U_M = NI$ is called "magnetic tension".

With the help of these analogies, magnetic fields can now mapped onto electric circuits. The circuit elements and their magnetic counterparts are sketched in ◘ Fig. 5.14. As a result, all strategies for the analysis of electrical networks are also available for the analysis of magnetic fields.

❗ Context Specific Terminology

The terms "magnetic voltage", "magnetic field strength", "magnetic resistance", and "magnetic flux density" are bound to the context of electrical machines and transformers. They have no physical meaning outside of this context.

Finally, it should be noted that everything that can be calculated with magnetic circuits can also be calculated without them. The advantage of mapping onto an electrical circuit arises if the magnetic field consists of many partial circuits. In this case, the techniques of mesh current analyses or the superposition method can also be used for magnetic fields.

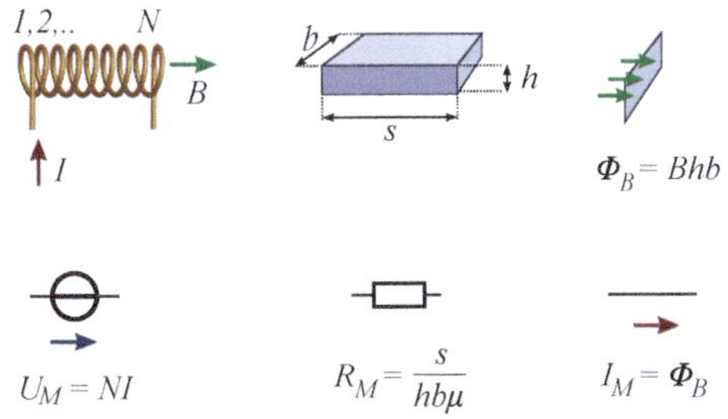

◘ **Fig. 5.14** Mapping magnetic circuits onto DC circuits: A current-carrying winding creates a magnetic voltage, a piece metal has a magnetic resistance. The magnetic flux is mapped onto the electrical current (graphics from [6])

$\Phi_B = Bhb$

$U_M = NI$

$R_M = \dfrac{s}{hb\mu}$

$I_M = \Phi_B$

5.8 **Problems**

5.1 What (useful) relationships can be derived from Gauss' law for an arbitrary vector field V, when applied to the special cases $V = f\,C$ and $V = F \times C$ (C is constant)? Please look for equations in which C no longer occurs.

5.2 Please express the path element ds' appearing in Biot-Savart's law and also showing up in the multipole expansion of the magnetic potential (see ■ Fig. 5.4 or 5.11) as a function of the vectors r and r'. Then, find a replacement for the integral $\oint_{\partial a} f\,ds'$ by using Stokes' theorem for the vector field $V = f\,C$ (with C being constant).

5.3 Show that the integral formula for determining the magnetic vector potential from the distribution of the current density J, ■ Eq. (5.12) reduces to a path integral along the wire (5.15) if the current is limited to a thin wire.

5.4 A current I flows in the rectangular loop, as shown in ■ Fig. 5.15, in a plane perpendicular to a magnetic field $B = (0, 0, B)$. The strength of the field varies as a function of the spatial variable x, only: $B = B(x)$. Please find an expression for the force acting on the conductor loop.

5.5 A rectangular conductor loop is placed in the magnetic field $B = (0, 0, 2)$ mT. The loop connects the points

$$P_1 = (-3, -1, -2) \text{ cm}, \quad P_2 = (3, -1, -2) \text{ cm},$$
$$P_3 = (3, 2, 4) \text{ cm}, \quad P_4 = (-3, 2, 4) \text{ cm}$$

■ **Fig. 5.15** Illustrating ■ Problem 5.4: A conductor loop carrying a current I which lies in a plane perpendicular to the magnetic field B. The loop passes through the points a, b, c and d

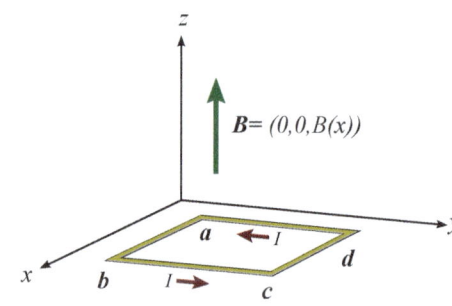

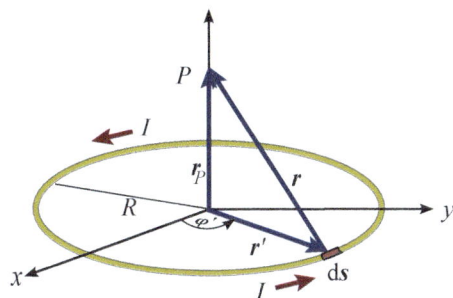

Fig. 5.16 Illustrating
Problem 5.6: A current
I flowing in a ring. The field
B at the point *P* is to be determined

in the order given. A current of $I = 5$ A flows in the loop. Please determine the torque vector and the axis around which the conductor loop is rotated.

5.6 A current I flows through a ring with radius R and negligible thickness (see ▪ Fig. 5.16). Please determine the field of the magnetic force **B** along its axis of symmetry. Hint: Use Biot-Savart's law and replace the piece d**s** by $(d\mathbf{r}'/d\phi')d\phi'$.

5.7 ▪ Figure 5.17 shows the cross-section through an electromagnet that carries a piece of iron. The area a across the field lines is constant everywhere. The magnetic field lines have a total length g in the iron and an additional small portion d on the left and right in the air. Determine the strength of the field B using Ampere's law. Derive the total energy of the magnetic field. Then, interpret this energy as the potential energy of the entire system. Using

$$W = \int F d\mathbf{x}, \text{ calculate the force acting on the piece of iron. Hint: } \int_{0}^{d} F d\mathbf{x}$$

Fig. 5.17 Illustrating
Problem 5.7: An electromagnet is powered by a
winding with N turns carrying
a current current I. It attracts
a piece of iron at a distance d
below

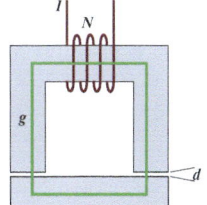

reduces to $-\int\limits_{0}^{d} F\mathrm{d}x$ in this case. Now, the derivative with respect to d can bet taken (fundamental theorem).

5.8 The preservation of the magnetic flux in iron structures is based on the fact that "magnetic field lines run almost parallel to the surface of the iron." Some books on engineering also postulate that "magnetic field lines leave the iron at an angle of 90°." Please justify the two statements by an analysis of the material combination $\mu_{Fe} = 1000$ and $\mu_{air} = 1$. Draw the angle α_{Fe} as a function of the angle α_{air} and the inverse function $\alpha_{air}(\alpha_{Fe})$. The angles are defined in ◘ Fig. 3.9.

5.9 ◘ Figure 5.18 shows the cross section through a coaxial cable. The conducting materials are limited by the radii R_1 and R_3 and separated by an insulator extending from R_1 to R_2. A current I_i flows through the inner conductor, and a current I_a flows through the outer conductor in the opposite direction. Please specify the field of the magnetic force (the "flux density") B inside and outside of the cable. For all materials, $\mu_{\mathrm{r}} \approx 1$ may be assumed.

5.10 ◘ Figure 5.19 shows a very thin metal strip that is placed on the x-axis. Its height h extends from $z = -h/2$ to $z = h/2$. Please determine the vector potential A and the field B for a point P on the y-axis. The length l is very

◘ **Fig. 5.18** Cross-section through a coaxial cable consisting of an inner conductor, an insulator and an outer conductor

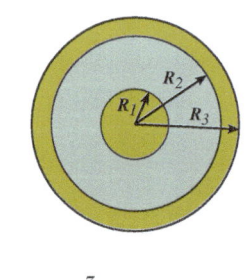

◘ **Fig. 5.19** A very narrow metal band of length l and height h which is of one Current I flows through in the x direction.

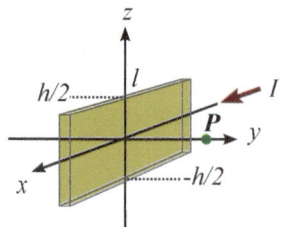

large. Hint: A good starting point for this is the vector potential for a long wire.

5.9 Solutions

5.1 The special cases of Gaus' theorem are for $V = f\,C$

$$\int_V \nabla \cdot (f\,C)\mathrm{d}V' = \int_V [(\nabla f) \cdot C + 0]\mathrm{d}V' = \left(\int_V \nabla f\,\mathrm{d}V'\right) \cdot C$$

$$\text{and} \quad \oint_{\partial V} f\,C \cdot \mathrm{d}a' = C \cdot \oint_{\partial V} f\,\mathrm{d}a'$$

$$\rightarrow \quad \int_V \nabla f\,\mathrm{d}V' = \oint_{\partial V} f\,\mathrm{d}a'$$

and for $V = F \times C$

$$\int_V \nabla \cdot (F \times C)\,\mathrm{d}V' = \int_V [C \cdot (\nabla \times F) - 0]\,\mathrm{d}V' = C \cdot \int_V \nabla \times F\,\mathrm{d}V'$$

$$\text{and} \quad \oint_{\partial V} (F \times C) \cdot \mathrm{d}a' = \oint_{\partial V} (\mathrm{d}a' \times F) \cdot C = C \cdot \oint_{\partial V} \mathrm{d}a' \times F$$

$$\rightarrow \quad \int_V \nabla \times F\,\mathrm{d}V' = \oint_{\partial V} \mathrm{d}a' \times F$$

5.2 Here, $\mathrm{d}s' = \mathrm{d}r'$. The special cases of Stokes theorem are

$$\int_a [\nabla \times (f\,C)] \cdot \mathrm{d}a' = \int_a [(\nabla f) \times C + 0] \cdot \mathrm{d}a' = C \cdot \int_a \mathrm{d}a' \times (\nabla f)$$

$$\text{and} \quad \oint_{\partial a} f\,C \cdot \mathrm{d}s' = C \cdot \oint_{\partial a} f\,\mathrm{d}s'$$

$$\rightarrow \quad \int_a \mathrm{d}a' \times (\nabla f) = \oint_{\partial a} f\,\mathrm{d}s'$$

5.3 As shown in ◘ Fig. 5.20, the vectors J, O and ds are parallel so that the direction of the final result is that of ds. A disk Ods serves as the volume element. The integration to determine the vector potential (5.12) can then be written as

$$A = \frac{\mu}{4\pi} \int_V \frac{J(r')}{|r - r'|} dV' = \frac{\mu}{4\pi} \int_S \left(\int_O \frac{J(r')}{|r - r'|} dO' \right) ds'.$$

If the wire is very thin, and if the field is to be determined at a distance much larger than the radius of the wire, the variation of $1/|r - r'|$ can be neglected when integrating over the cross-sectional area O. This factor can, therefore, be placed before the integral over the surface O. Then,

$$A = \frac{\mu}{4\pi} \int_S \left(\frac{1}{|r - r'|} \int_O J(r') dO' \right) ds',$$

follows. The integral over O' determines the current passing through the wire. Inserting this current gives ◘ Eq. (5.15).

5.4 Using ◘ Eq. (5.17) one can write

$$F = I \oint (ds \times B) = I \oint \begin{vmatrix} e_x & e_y & e_z \\ dx & dy & 0 \\ 0 & 0 & B \end{vmatrix} = I \oint (dyB, -dxB, 0),$$

because, in this case, the pieces of the path are parallel to the axes. The path integral is now divided into four sections. When writing it down explicitly,

$$F = I \left[\int_a^b (0, -Bdx, 0) + B(b) \int_b^c (dy, 0, 0) + \int_c^d (0, -Bdx, 0) + B(a) \int_d^a (dy, 0, 0) \right]$$

◘ **Fig. 5.20** Illustrating ◘ Problem 5.3: The parallel vectors current density J, area O and line segment ds of a thin wire

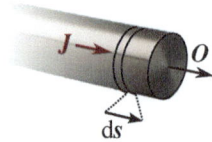

it is taken into account that the magnetic field does not change along the y-axis. The first and third integrals differ only in the sign. Their sum is zero. The same holds for the second and fourth integrals, except for them having different pre-factors of B. This leaves

$$F = I(B(b) - B(a)) \left[\left(\int_b^c dy \right), 0, 0 \right] = I\left[(B(b) - B(a))(c - b), 0, 0 \right]$$

as the solution. If the current reverses direction, so does the force. If the field is homogeneous, no force acts---as expected.

5.5 Before applying the formula $\tau = a \times B$, it is checked whether the points form a rectangle. To do this, one can calculate the distance vectors $R_{ik} = P_k - P_i$ so that their direction corresponds to that of the current:

$$R_{12} = (6, 0, 0)\ \text{cm} \qquad R_{23} = (0, 3, 6)\ \text{cm}$$
$$R_{34} = (-6, 0, 0)\ \text{cm} \quad R_{41} = (0, -3, -6)\ \text{cm}$$

These vectors are pairwise antiparallel, and two consecutive vectors are perpendicular. So, the vectors form a rectangle. The area vector a can be thus calculated as any of the four vector products

$$a = R_{12} \times R_{23} = R_{23} \times R_{34} = R_{34} \times R_{41} = R_{41} \times R_{12} = (0, -36, 18)\ \text{cm}^2$$

and the torque is

$$\tau = a \times B = (-72, 0, 0)\ \text{cm}^2\ \text{mT} = (-1, 0, 0) \cdot 7.2\ \mu\text{Nm}.$$

The axis of rotation must pass through the center of the rectangle $(P_1 + P_3)/2 = (P_2 + P_4)/2 = (0, \frac{1}{2}, 1)$, and it must be parallel to the x-axis. So $(x, \frac{1}{2}, 1)$ describes the axis.

5.6 This problem can be solved using the Biot-Savart law. ◼ Figure 5.16 shows the variables required to solve the problem. The setup is symmetric around the axis of the ring. So, the variables are first determined as a function of the angle ϕ' in the xy plane. The final result is then obtained by integration over ϕ'. The terms needed for an application of the Biot-Savart law are then: The vector to a point on the ring is $r' = R\left(\cos\phi', \sin\phi', 0\right)$.

The chain rule is used to determine the line element ds'[6]:

$$I ds' = I \frac{\partial r'}{\partial \phi'} d\phi' = IR d\phi' (-\sin\phi', \cos\phi', 0)$$

The vector between the ring elements and the point $r_p = (0, 0, h)$ at which the magnetic field is to be determined is $r = r_p - r' = (-R\cos\phi', -R\sin\phi', h)$. Its length is $|r| = \sqrt{h^2 + R^2}$.

Now $I ds' \times r$ can be determined using the unit vectors e_x etc. to be

$$I ds' \times r = \begin{vmatrix} e_x & e_y & e_z \\ -\sin\phi' & \cos\phi' & 0 \\ -R\cos\phi' & -R\sin\phi' & h \end{vmatrix} IR d\phi' = (h\cos\phi', h\sin\phi', R) IR d\phi',$$

which finally leads to an application of the law:

$$B = \frac{\mu IR}{4\pi \left(R^2 + h^2\right)^{3/2}} \int_0^{2\pi} (h\cos\phi', h\sin\phi', R) \, d\phi'.$$

With $B = (B_x, B_y, B_z)$, the integration gives $B_x = B_y = 0$ and

$$B_z = \frac{\mu IR^2}{2(R^2 + h^2)^{3/2}}.$$

5.7 Conservation of the magnetic flux implies that a constant cross-section goes along with a constant field strength B. Hence,

$$NI = \oint \frac{B}{\mu} dx \approx \frac{Bg}{\mu_0 \mu_{Fe}} + \frac{2Bd}{\mu_0} = \frac{B}{\mu_0} \left(\frac{g}{\mu_{Fe}} + 2d \right)$$

is a good approximation. The energy density of the magnetic field is

$$\frac{B^2}{2\mu_0 \mu_r} = \frac{W}{V} \rightarrow W = \frac{B^2}{\mu_0} a \left(\frac{g}{2\mu_{Fe}} + d \right).$$

Now, B can be eliminated:

---

6   This takes advantage of the fact that the other partial derivatives disappear.

$$W = \frac{\mu_0 (NI)^2 a}{2 \left(\frac{g}{\mu_{Fe}} + 2d \right)}$$

The fundamental theorem of integration

$$W = - \int_0^d F \, dx \rightarrow F = -\frac{\partial W}{\partial d}.$$

can be used to determine the force.[7] Consequently,

$$F = -\frac{\partial W}{\partial d} = \frac{\mu_0 (NI)^2 a}{\left(\frac{g}{\mu_{Fe}} + 2d \right)^2} = \frac{B^2 a}{\mu_0}$$

is the force that was sought. For the final transformation the first equation of this problem solution was used. The term in the denominator explains why pieces of iron stick to electromagnets as if they were glued: For $d = 0$, the force is proportional to μ_{Fe}. For large distances, μ_{Fe} becomes irrelevant. For electromagnets that are significantly smaller than 1 m, the attraction is reduced by more than a factor of 1000 within the distance of one millimetre.

5.8 According to ◘ Eq. (3.27), the angle to the surface normal changes between "medium 1" and "medium 2" in the following manner:

$$\alpha_2 = \arctan \left[\frac{\mu_2}{\mu_1} \tan(\alpha_1) \right]$$

If one sets $\mu_1 = \mu_{air} = 1$ and $\mu_2 = \mu_{Fe} = 1000$, the graph shown in in ◘ Fig. 5.21 results. The fact that, within the iron, the angle is close $\alpha = 90°$ almost everywhere is the origin of magnetic flux conservation in iron.

5.9 Ampere's law (3.25) applies to all areas. Furthermore, the current densities are constant within regions of any specific material. Then,

---

7   In this case, the partial derivative suffices, since all other parameters are constant.

Fig. 5.21 Illustrating Problem 5.8: Angular dependencies of magnetic field lines at the surface of iron in air. If the line leaves the iron (blue line), the angle there is close to 0° almost everywhere. The field line is perpendicular to the surface. A field line entering the iron continues nearly parallel to the surface. ($\alpha \approx 90°$)

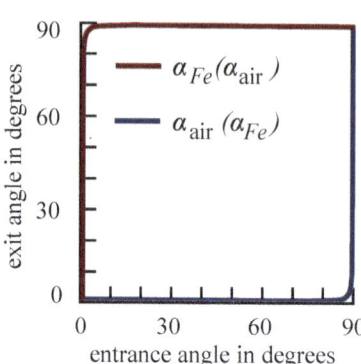

angle between the field lines and the surface normal

range $r < R_1$: $I = I_i \dfrac{r^2}{R_i^2}$ $\rightarrow$ $B = \mu_0 I_i \dfrac{r}{2\pi R_1^2}$

range $R_1 < r < R_2$: $I = I_i$ $\rightarrow$ $B = \mu_0 I_i / (2\pi r)$

range $R_2 < r < R_3$: $I = I_i - I_a \left(\dfrac{r^2 - R_2^2}{R_3^2 - R_2^2} \right)$ $\rightarrow$ $B = \dfrac{\mu_0}{2\pi r} \left[I_i - I_a \left(\dfrac{r^2 - R_2^2}{R_3^2 - R_2^2} \right) \right]$

range $R_3 < r$: $I = I_i - I_a$ $\rightarrow$ $B = \mu_0 (I_i - I_a) / (2\pi r)$

are the results.

5.10 Since the current flows in the x direction, the vector potential has one component only: $A = (A, 0, 0)$. For a very long wire, one gets

$$A = \frac{\mu}{2\pi} I \cdot \ln\left(\frac{2l}{r}\right) \quad \text{(very thin, very long wire)},$$

where r is the distance to the wire.

If many wires are placed on top of each other, the configuration shown in **Fig. 5.19** is created. Mathematically, this means for a point $P = (0, y, 0)$

$$A = \frac{\mu}{2\pi} \sum_{k=1}^{N} I_k \cdot \ln\left(\frac{2l}{\sqrt{y^2 + z_k^2}}\right) \qquad \text{(discrete with } I_k = I/N)$$

$$\rightarrow A = \frac{\mu}{2\pi} \int_{-h/2}^{h/2} \left(\frac{dI}{dz'}\right) \cdot \ln\left(\frac{2l}{\sqrt{y^2 + z'^2}}\right) dz' \quad \left(\text{continuous with } \frac{dI}{dz'} = \frac{I}{h} \right)$$

Fig. 5.22 The ratio of the
field strength B of a narrow
metal strip to that of a very
thin wire as a function of the
distance y normalized to the
band height h

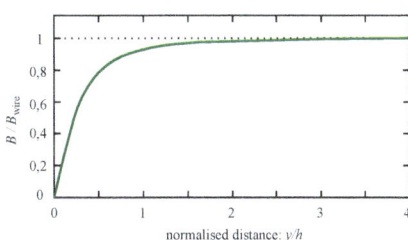

and, via integration, leads to the first solution :

$$A = \frac{\mu}{2\pi} \cdot I \cdot \left\{ \ln \left(\frac{2l}{\sqrt{4h^2 + y^2}} \right) + 1 - \frac{2y}{h} \cdot \arctan \left(\frac{h}{2y} \right) \right\} \qquad (5.32)$$

The field $\mathbf{B}$ can have a component in the z-direction, only, $\mathbf{B} = (0, 0, B)$. In this case, the equation $\mathbf{B} = \nabla \times \mathbf{A}$ is reduced to $B = -\partial A / \partial y$. After several cancellations,

$$B = \frac{\mu I}{\pi h} \cdot \arctan \left(\frac{h}{2y} \right) \qquad (5.33)$$

emerges which approaches the formula for the field of a thin wire for $h \to 0$ - as required. It may be striking that the field is free of divergences for $h \neq 0$:

$$B(y = 0) = \frac{\mu I}{2h} .$$

The comparison with the field of the thin wire is also interesting. If B is written as the product of the wire field and a modification function $F = B / B_{\text{wire}}$ it can be written as

$$B = B_{\text{wire}} \cdot F(\alpha) = \frac{\mu I}{2\pi y} \cdot \left(2\alpha \arctan \frac{1}{2\alpha} \right)$$

with $\alpha = y/h$. ▪ Figure 5.22 shows that the field strength approaches that of a thin wire if y/h exceeds the value 1.

References

1. Emil Warburg, Annalen der Physik und Chemie 20, S. 814-835, Leipzig 1881
2. Dong Lin und Xin Chen, Mathematical Models of 3D Magneti Field and 3D Positioning System by Magnetic Field, Appl. Math. Inf. Sci. 8 No 4, 2014, P1647-1654., Natural Science Publishing Cor. ► www.naturalspublishing.com/files/published/54ims25g64sp5o.pdf
3. Marco Leone, Theoretische Elektrotechnik, 2.Auflage, Springer Heidelberg 2021, ISBN 978-3658292072
4. Gottlieb und Peter Strassacker, Analytische und numerische Methoden der Feldberechnung, Vieweg und Teubner 1993, ISBN 978-3-519-06168-7
5. Arnulf Kost, Numerische Methoden in der Berechnung elektromagnetischer Felder, Springer Heidelberg 1994, ISBN 978-3-540-55005-1
6. Martin Poppe, Pruefungstrainer Elektrotechnik, 4. Auflage, Springer Heidelberg 2022, ISBN 978-3-662-65001-1

5

Some Properties of Time-Dependent Fields

Contents

© The Author(s), under exclusive license to Springer-Verlag GmbH,
DE, part of Springer Nature 2024
M. Poppe, *Basic Electrodynamics in 6 Lessons*,
https://doi.org/10.1007/978-3-662-69143-4_6

Abstract The effort required to calculate time-dependent systems is significantly larger than for calculating static systems. So, everything has to be done to keep the complexity within tolerable limits. First, a decoupled scalar differential equations system will be introduced. These equations deal with potentials. The decoupling of space and time components is done using a method known as gauging. The equations obtained in this way determine the behaviour of electromagnetic waves in space and at material boundaries. As a result, the laws of optics emerge. It is also shown that time-dependent fields always carry energy and momentum---sometimes in a counter-intuitive way. Limits for simplifications by using quasi-static approximations are discussed. The chapter ends with a description of the electrical behaviour of cables.

6

6.1 Electrodynamics: A Gauge Field Theory

The fields E and B can be calculated separately in the static case. Once they become time-dependent, they also become dependent on each other, and the mathematical effort increases substantially. The Ampere-Maxwell law and the Faraday-Henry law define their interdependence. These laws form the basis to describe electromagnetism using potentials. The result allows us to map the four coupled Maxwell's equations onto a system of two separate equations, but at a price: These differential equations are second order. Further simplifications may be achieved by a method called "gauging".

The Vector potential A in ◘ Eq. (5.6) implies Gauss' law for the magnetic field because **div (rot V)** $= 0$ is valid for every vector V. Using the Faraday-Henry law, one can create a potential for the electric field, which is a pure source field even in the dynamic case, by inserting A in a clever manner. The transformations

$$B = \nabla \times A \mid \text{definition of } A$$

$$\text{and} \quad \nabla \times E + \tfrac{\partial B}{\partial t} = 0 \qquad \mid \text{Faraday} - \text{Henry law}$$

$$\rightarrow \nabla \times \left(E + \tfrac{\partial A}{\partial t}\right) = 0 \qquad \mid \text{inserted}$$

$$\rightarrow \qquad E + \tfrac{\partial A}{\partial t} = -\nabla\Phi \quad \mid (E + \partial A/\partial t) \text{ must be a source field}$$

lead to the generalisation of ◘ Eq. (4.6). Not only are the fields E and B defined by the potentials A and Φ via

$$B = \nabla \times A \quad \text{and} \quad E = -\nabla\Phi - \frac{\partial A}{\partial t} \tag{6.1}$$

also, they a priori satisfy Gaus' theorem for the field B and the Faraday-Henry's law.

Next, it must be checked whether the two remaining Maxwell equations are fulfilled as well. By inserting ▢ Eq. (6.1) into ▢ Eq. (3.12), i.e.

$$\nabla \cdot \left[\varepsilon \left(\nabla\Phi + \frac{\partial A}{\partial t} \right) \right] = \rho_{\text{free}}$$

$$\nabla \times \left[\frac{1}{\mu} (\nabla \times A) \right] + \varepsilon \frac{\partial}{\partial t} \left(\nabla\Phi + \frac{\partial A}{\partial t} \right) = J_{\text{free}} \tag{6.2}$$

one gets the desired result: A set of potential equations which are valid in any material. If ε and μ are constant numbers, the equations simplify because the rotation terms can be eliminated. With $v = 1/\sqrt{\varepsilon\mu}$ they read

$$\nabla \cdot (\nabla\Phi) + \frac{\partial}{\partial t}(\nabla \cdot A) = -\rho_{\text{free}}/\varepsilon$$

$$\nabla^2 A - \frac{1}{v^2} \frac{\partial^2 A}{\partial t^2} - \nabla \left(\nabla \cdot A + \frac{1}{v^2} \frac{\partial\Phi}{\partial t} \right) = -\mu J_{\text{free}} . \tag{6.3}$$

In a vacuum, v can be replaced by the speed of light c (more on the meaning of v in the context of the "Lorentz gauge" below). If no distinction is made between free and bound charges, the indices "free" are omitted. Then, ε_0 and μ_0 should be used in ▢ Eq. (6.3) instead of ε and μ.

Potentials that fulfil the systems of ▢ Eqs. (6.2) or (6.3) automatically lead to fields which are consistent with all four Maxwell equations. For the sake of simplicity, constant values for ε and μ will be assumed for the rest of this chapter so ▢ Eq. (6.3) will suffice from now on.

The potentials Φ and A can be modified within a specific framework without changing the fields E and B. Such modifications are known as *gauge transformations*. These transformations are widespread because they can simplify the equations for the potentials. The gauge transformations are

$$A_\Psi = A + \nabla\Psi$$

$$\Phi_\Psi = \Phi - \partial\Psi/\partial t , \tag{6.4}$$

where Ψ is a scalar function of r and t. The field B does not change as a result of this transformation, i.e. $B_\Psi = B$. This is a consequence of ▢ Eq. (6.1) because $\nabla \times (\nabla\Psi) = 0$. To show that E does not change either, the following constructive path is used [1]: Instead of asking whether the transformation

(6.4) is correct, one checks: Given the transformation for A, what kind of transformation of Φ is required to keep electric field E unchanged? In other words, What is Φ_Ψ? The answer is obtained by inserting the transformed potential into the equation giving E:

$$E_\Psi = -\tfrac{\partial}{\partial t} A_\Psi - \nabla\Phi_\Psi \qquad |\text{ insert}$$

$$= -\tfrac{\partial}{\partial t}(A + \nabla\Psi) - \nabla\Phi_\Psi \quad|\text{ collect divergences}$$

$$= -\tfrac{\partial}{\partial t}A - \nabla\left(\tfrac{\partial}{\partial t}\Psi + \Phi_\Psi\right) |\text{ compare with original equation}$$

$$E = -\tfrac{\partial}{\partial t}A - \nabla\Phi \qquad\qquad |\text{ Eq. (6.4) is fulfilled.}$$

The comparison of the last two lines shows that the electric field does not change under the gauge transformation, so $E = E_\Psi$.

The freedom offered by a gauge transformation can be used to reduce computing efforts. The two options, *Coulomb gauge* and *Lorentz gauge* are the most popular ones. The Lorentz gauge will be introduced briefly. From then on, the Lorentz gauge will be used.

The Coulomb gauge shortens calculations if $\rho = 0$

The Coulomb gauge is defined by

$$\nabla \cdot A = 0 \quad \text{(Coulomb gauge)}$$

This gauge is also used for static magnetic fields, and when charges average out to zero.

According to ◾ Eq. (6.3) the potentials in the Coulomb gauge fulfil

$$\nabla^2\Phi = -\frac{\rho_{\text{free}}}{\varepsilon} \quad \text{and} \quad \nabla^2 A - \frac{1}{v^2}\left(\frac{\partial^2 A}{\partial t^2} + \frac{\partial^2 \Phi}{\partial t^2}\right) = -\mu J_{\text{free}} \qquad (6.5)$$

and thus pave the way to determine Φ by integration: The potential Φ is a solution of the Poisson equation (4.9) that can be solved using ◾ Eq. (4.15).

The Coulomb gauge also often leads to complicated systems to solve because both potentials, Φ and A, appear in a single equation. Therefore, the Lorentz gauge will be used for the rest of the book.

The Lorentz gauge separates potentials

The full power of gauge transformations is revealed when Ψ is chosen so that

$$\nabla \cdot A + \frac{1}{v^2} \frac{\partial \Phi}{\partial t} = 0 \quad \text{(Lorentz gauge)} \tag{6.6}$$

is fulfilled. Then, the ■ Eq. (6.3) simplify and decouple:

$$
\begin{aligned}
\nabla^2 \Phi - \frac{1}{v^2} \frac{\partial^2 \Phi}{\partial t^2} &= -\rho_{\text{free}}/\varepsilon \\
\nabla^2 A - \frac{1}{v^2} \frac{\partial^2 A}{\partial t^2} &= -\mu J_{\text{free}} .
\end{aligned}
\tag{6.7}
$$

In the absence of matter

$$
\begin{aligned}
\nabla^2 \Phi - \frac{1}{c^2} \frac{\partial^2 \Phi}{\partial t^2} &= -\rho/\varepsilon_0 \\
\nabla^2 A - \frac{1}{c^2} \frac{\partial^2 A}{\partial t^2} &= -\mu_0 J
\end{aligned}
$$

applies. Equations of type (6.7) are called *inhomogeneous wave equations* because their solutions describe waves that move with the velocity

$$
\begin{aligned}
v &= \frac{1}{\sqrt{\mu\varepsilon}} \quad \text{in matter and} \\
c &= \frac{1}{\sqrt{\mu_0\varepsilon_0}} \quad \text{in a vacuum.}
\end{aligned}
\tag{6.8}
$$

Since light is an electromagnetic wave, c is known to be the speed of light.

Even in the case of time-dependent fields, one can work with four independent equations (the components of A are not mixed). Every solution of this system of equations, together with the gauge condition (6.6), necessarily leads to fields E and B, which fulfil all four Maxwell equations. Therefore, as long as μ and ε are constant, one can conclude:

❯ In the Lorentz gauge, equations decouple
The Lorentz gauge reduces the determination of dynamic electromagnetic fields to the solution of four independent, inhomogeneous wave equations.

In the absence of charges and currents the system (6.7) is reduced to the *homogeneous wave equations*

■ **Fig. 6.1** Wave propagation with speed v in a pond from point r as a result a stone's throw. The outer wave passes through the points $r_1 = r + v_1 t$ and $r_2 = r + v_2 t$

$$\nabla^2 \Phi - \frac{1}{v^2} \frac{\partial^2 \Phi}{\partial t^2} = 0$$
$$\nabla^2 A - \frac{1}{v^2} \frac{\partial^2 A}{\partial t^2} = 0 \, .$$

(6.9)

which will be investigated first.

Only one of d'Alembert's solutions is consistent with causality

The general solution to the wave equation is named after its discoverer. It is called *d'Alembert's solution*: If and only if waves do not separately depend on the location r and time t, but on the combination $r \pm vt$, they are a solution of ■ Eq. (6.9). This dependency may be observed by throwing a little stone into a pond (see ■ Fig. 6.1).

So, the equations should look like

$$A = A_1(r - vt) + A_2(r + vt) \quad \text{and}$$
$$\Phi = \Phi_1(r - vt) + \Phi_2(r + vt) \, .$$

(6.10)

The system (6.10) consists of four independent equations, in Cartesian coordinates, for example for A_x, A_y, A_z and Φ. To show that ■ Eq. (6.10) solves the wave equation, the second derivatives of the potential $\Phi_1(r - vt)$ shall be calculated and inserted into ■ Eq. (6.9). The calculation for Φ_2 and for the components of A proceed analogously.

According to the chain rule,[1] the second spatial derivative can be calculated like this,

---

1   The following, $\partial \Phi_1 / \partial(r - vt)$ is the single-row Jacobian matrix $\left[\partial \Phi_1 / \partial(x - v_x t), (\partial \Phi_1 / \partial(y - v_y t), (\partial \Phi_1 / \partial(z - v_z t) \right]$. The second derivative is a three-by-three matrix. But these details are irrelevant because, in the end, these terms vanish by cancellation.

$$\frac{\partial \Phi_1}{\partial x} = \frac{\partial \Phi_1}{\partial (r - vt)} \frac{\partial (r - vt)}{\partial x} = \frac{\partial \Phi_1}{\partial (r - vt)} \cdot \boldsymbol{e}_x$$

$$\rightarrow \frac{\partial^2 \Phi_1}{(\partial x)^2} = \frac{\partial^2 \Phi_1}{[\partial (r - vt)]^2} \frac{\partial (r - vt)}{\partial x} \cdot \boldsymbol{e}_x = \frac{\partial^2 \Phi_1}{[\partial (r - vt)]^2}$$

while for the second timely derivative, one gets

$$\frac{\partial^2 \Phi_1}{(\partial t)^2} = \frac{\partial^2 \Phi_1}{[\partial (r - vt)]^2} \frac{\partial (r - vt)}{\partial t} \cdot (-v) = \frac{\partial^2 \Phi_1}{[\partial (r - vt)]^2} v^2,$$

so that ◻ Eq. (6.9) is fulfilled.

D'Alembert's formula defines a set of solutions rather than a single one. This is a consequence of the fact that it does not specify *how* Φ depends on $(r \pm vt)$, just *that* it does. At the same time, it distinguishes between the physically sensible solutions and the others: Just like the spread of the waves in a pond ◻ Fig. 6.1 follows the stone's throw temporally and causally (no stone, no wave), electromagnetic fields follow their cause with a time delay given by the distance and the speed of light.

Fields from charges are delayed

Time-dependent electrical potentials can now be determined analogously to that of static potentials: First, the potential of a point-like charge is determined, then the result is generalised to a set of charges. Finally, equations for continuous distributions can be found.

If there is only one point-like charge carrier at the coordinate origin $(x, y, z) = (0, 0, 0)$, the homogeneous wave equation applies everywhere except at this point. In addition, the potential must have a spherically symmetrical shape. Then, the potential can only depend on the radius $r = \sqrt{x^2 + y^2 + z^2}$. In mathematical terms, this means.[2]

$$\nabla^2 \Phi_{\text{point−like charge}} = \Delta \Phi_{\text{point−like charge}} = \frac{1}{r} \frac{\partial^2}{\partial r^2} \left(r \Phi_{\text{point−like charge}} \right).$$

So, except for the origin, the wave equation (6.9) may be written as

$$\frac{\partial^2}{\partial r^2} \left(r \Phi_{\text{point−like charge}} \right) - \frac{1}{v^2} \frac{\partial^2}{\partial t^2} \left(r \Phi_{\text{point−like charge}} \right) = 0.$$

---

2   Here, $\partial^2 (r\Phi)/\partial r^2 = \partial [r^2 \partial \Phi / \partial r]/\partial r$ is used.

Here, too, the solution must follow d'Alembert, in this case, with a minus sign in front of the vt term. Which sign is physically meaningful depends on the formulation of the problem:

1. Question about the **effect**: Is the time of the cause fix? (e.g. charge at a location r at time $t = 0$), then, solutions of the type $\Phi_2(r + vt)$ make sense.
2. Question about the **cause**: Is the time of the effect fix? (e.g. potential at a location r_1 at a time $t = 0$), then, the solution for the charge is of the type $Q = Q(r_1 - vt)$.

A second look at ◘ Fig. 6.1 confirms the correct sign when choosing the solution: The outer wave is at the points $r_i = r + v_i t$, where t is the time between the impact of the stone and the moment when the picture was taken. If one wants to deduce the location of the impact of the stone from the outer wave at location r, then, $r = r_i - v_i t$ is the correct solution.

Now, one can seek a potential of the type

$$\Phi_{\text{point-like charge}}(r, t) = \frac{1}{r} f(r - vt) \rightarrow \Phi_{\text{point-like charge}}(r, t) = \frac{1}{r} g\left(t - \frac{r}{v}\right).$$

The function g exploits the spherical symmetry of the problem. It is determined by a comparison with the static potential (4.13) because for $v \rightarrow \infty$, both must agree at any time t. The potential

$$\Phi_{\text{point-like charge}}(r, t) = \frac{Q\left(t - \frac{r}{v}\right)}{4\pi \varepsilon r} \tag{6.11}$$

fulfils this requirement. The expression (6.11) is know as the *retarded potential* or *delayed potential* of the charge Q. It implies that at a distance r, the effect of a charge comes with a delay $\Delta t = r/v$.

The proof of $\Phi_{\text{point-like charge}}$ being the correct potential function is done by showing that everywhere except at the point $(0, 0, 0)$, the homogeneous wave equation (6.9) is fulfilled:

Since the position of the charge was chosen to be the coordinate origin, $\partial/\partial r$ is the only relevant component of the Nabla operator. Consequently, the validity of

$$\Delta \Phi_{\text{point-like charge}} = \frac{1}{r^2} \frac{\partial}{\partial r} r^2 \frac{\partial}{\partial r} \Phi_{\text{point-like charge}} = \frac{1}{v^2} \frac{\partial^2}{\partial t^2} \Phi_{\text{point-like charge}}$$

suffices to show that the potential function is correct. In this case, it helps to take the derivatives concerning the *reduced time* $\tau_- = t - r/v$, using the chain rule

$$\frac{\partial Q}{\partial t} = \frac{\partial Q}{\partial \tau_-}\frac{\partial \tau_-}{\partial t} = \frac{\partial Q}{\partial \tau_-} \quad, \quad \frac{\partial \tau_-}{\partial r} = -\frac{1}{v} \quad \text{and} \quad \frac{\partial(1/r)}{\partial \tau_-} = \frac{v}{r^2} \quad,$$

and the second spatial derivative then comes out as

$$\Delta \Phi_{\text{pointcharge}} = \frac{1}{4\pi \varepsilon r v^2}\frac{\partial^2 Q}{\partial t^2}$$

The second timely derivative differs from the spatial derivative by a factor v^2, only. Thus, ☐ Eq. (6.11) is correct.

The potential at any point r of a system with n point charges at the locations r_i can now be written down as the sum of individual potentials,

$$\Phi(r, t) = \frac{1}{4\pi \varepsilon} \sum_{i=1}^{n} \frac{Q_{i,\text{free}}\left(t - \frac{|r-r_i|}{v}\right)}{|r - r_i|} \tag{6.12}$$

and, in analogy to ☐ Eq. (4.14, 4.15), it can be generalised to continuous charge distributions,

$$\Phi(r, t) = \frac{1}{4\pi \varepsilon} \int_V \frac{\rho_{\text{free}}\left(t - \frac{|r-r'|}{v}\right)}{|r - r'|} \, dV', \tag{6.13}$$

where the apostrophes (') indicate the quantities to be integrated.

All components of the vector potential A are subject to the same differential equations. So they must also look like ☐ Eq. (6.13). Their normalisation can be derived from the fact that in the absence of charges and for $v \to \infty$, the potential should reproduce ☐ Eq. (5.12). The equation

$$A(r, t) = \frac{\mu}{4\pi} \int_V \frac{J_{\text{free}}\left(t - \frac{|r-r'|}{v}\right)}{|r - r'|} \, dV' \tag{6.14}$$

is the result of this normalisation.

In the absence of matter, μ and ε can be replaced by μ_0 and ε_0. In addition, the index "free" should be omitted and the speed v is the (vacuum) speed of light, c.

6.2 Waves at Material Boundaries: The Foundation of Optics

Reflection and refraction are phenomena known from optics. The following paragraphs will describe them as a consequence of the laws of electrodynamics. They turn out to be consequences of Maxwell's equations applied material interfaces.

Harmonic solutions can simplify calculations

Waves are called harmonic if their progression is sinusoidal in space as well as in time. In this chapter, the complex exponential function shall serve as an extension of the sine function for waves of the electric field, as it is done for describing alternating currents. So, the notation

$$\underline{E} = \underline{\hat{E}} e^{\mathrm{j}(\omega t - k \cdot r)} \tag{6.15}$$

shall be used.[3] The measurable electric wave $E(t, r)$ is then given by

$$E = \Im(\underline{E}) = \hat{E}\sin(\omega t - k \cdot r + \phi_E) \quad \text{with} \quad \underline{\hat{E}} = \hat{E}e^{\mathrm{j}\phi_E}, \tag{6.16}$$

i.e. it is the *imaginary* part of the complex wave function. Complex potentials of the form

$$\underline{\Phi} = \underline{\hat{\Phi}} e^{\mathrm{j}(\omega t - k \cdot r)} \quad \text{and} \quad \underline{A} = \underline{\hat{A}} e^{\mathrm{j}(\omega t - k \cdot r)} \tag{6.17}$$

will be assumed to determine the properties of harmonic electromagnetic waves. Inserting them into the wave equations (6.9) shows, that these functions are solutions if and only if the equation

$$k^2 v^2 = \omega^2 \tag{6.18}$$

is fulfilled, and its three variables are constant.

---

3   Complex variables are marked with an underscore. The complex amplitude $\underline{\hat{E}}$ includes the phase shift. For $\phi_E = 0$ the complex amplitude is real.

To determine the fields $\underline{E}$ and $\underline{B}$, a coordinate system is chosen in which the wave vector has the form $k = (k, 0, 0)$. Then, taking into account the gauge conditions (6.6), the fields

$$\underline{B} = \nabla \times \underline{A} \qquad = jk(0, \underline{A}_z, -\underline{A}_y) \qquad \text{and}$$
$$\underline{E} = -\nabla \underline{\Phi} - \tfrac{\partial}{\partial t}\underline{A} = -j\omega(0, \underline{A}_y, \underline{A}_z) \, , \qquad (6.19)$$

emerge. They are linked in various ways. And these links have implications for a travelling wave:

> In electromagnetic waves, the fields E, B and the wave vector k are perpendicular to one another. The two field vectors oscillate synchronously. The ratio of the amplitudes is $\hat{E}/\hat{B} = v$. The fields E and B described by ▢ Eq. (6.16) are sketched in ▢ Fig. 6.2 for $\phi_E = 0$.

Electric and magnetic waves are reflected differently

▢ Figure 6.3 shows the vector k for the reflection of an electromagnetic wave. It shall be assumed that the wave does not lose any energy, so there is a *total reflection* of the wave. The wave stays within one medium. So it retains its speed and its wavelength. The laws of reflection may either be traced back to the principle of minimal duration (the *Fermat principle*) or to the fact that

▢ **Fig. 6.2** One period of a transverse electromagnetic wave with $\phi_E = 0$. It propagates in the x direction ($k = (k, 0, 0)$). The arrows indicate the amplitudes of the fields

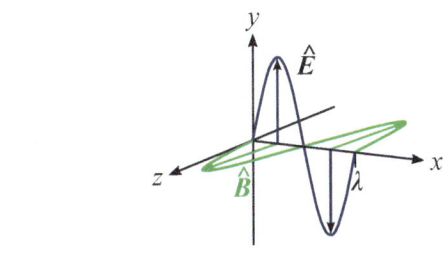

▢ **Fig. 6.3** On the geometry of the total reflection of an electromagnetic wave: a is the normal vector of the interface, k_I and k_R the wave vectors

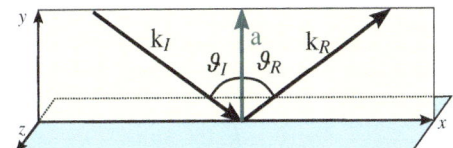

the ratio of the electric field strengths the incoming and outgoing waves must be the same at any time and at any point on the interface, as can be shown by the following reasoning:

Writing the incoming wave $\underline{E}_I$ and the reflected wave $\underline{E}_R$ as

$$\underline{E}_I = \hat{\underline{E}}_I e^{j(\omega_I t - k_I \cdot r)} \quad \text{and} \quad \underline{E}_R = \hat{\underline{E}}_R e^{j(\omega_R t - k_R \cdot r)},$$

gives the following ratio of incoming and outgoing waves

$$\frac{\underline{E}_I}{\underline{E}_R} = \frac{\hat{\underline{E}}_I}{\hat{\underline{E}}_R} e^{j[(\omega_I - \omega_R)t - (k_I - k_R) \cdot r]} .$$

If this ratio is to be correct for arbitrary times t, then $\omega_I = \omega_R$ must be true, i.e. the frequency does not change. If the ratio is to be the same for all points r_G on the surface, the equation $(k_I - k_R) \cdot r_G = 0$ must also be fulfilled. That means the difference between the wave vectors is perpendicular to the surface. Therefore, it is parallel to the surface normal vector a, i.e.

$$k_I - k_R \sim a .$$

If the surface is placed into the zx plane, the normal vector is $a = (0, a, 0)$, then the difference between the wave vectors must be proportional to the surface normal vector. Denoting the angle between the surface vector and a wave vector ϑ, the proportionality reads

$$k_I - k_R = k_I(\sin\vartheta_I, \cos\vartheta_I, 0) - k_R(\sin\vartheta_R, \cos\vartheta_R, 0) \sim (0, a, 0) .$$

This implies that the equations

$$k_I = k_R \quad \text{and} \quad \vartheta_I = \vartheta_R \tag{6.20}$$

are both fulfilled. ◼ Equation (6.20) are the *laws of reflection*.

The reflection laws do not give information about the polarisation of the incoming and outgoing field vectors. The field components in the plane of the surface are of particular interest. At normal incidence, the electric and magnetic field vectors are parallel to the surface, $E = E_\parallel$, $B = B_\parallel$. In the above coordinate system, the wave vectors are

$$k_I = (0, -k, 0) \quad \text{and} \quad k_R = (0, k, 0).$$

For a wave polarised in the z-direction,

$$\underline{E}_I = \underline{\hat{E}}_I e^{j(\omega t - k_I \cdot r)} = \hat{E}_I(0, 0, 1)e^{j(\omega t + ky)}$$

the Faraday-Henry law applied to the vicinity of the surface leads to the condition

$$\oint \boldsymbol{E} d\boldsymbol{l} = 0 \rightarrow E_{zI} + E_{zR} = 0 \rightarrow \hat{E}_R = -\hat{E}_I .$$

So, under reflection, the sign of the electrical field vector flips. The magnetic field shows a different behaviour: Using $\boldsymbol{B} = \boldsymbol{e}_k \times \boldsymbol{E}/v$, the magnetic wave can be deduced from the electric wave. The result

$$\boldsymbol{B}_I = \boldsymbol{B}_R = -\frac{E_I}{v}(1, 0, 0)e^{j(\omega t + ky)}$$

differs in sign from that of electric wave as shown in ◘ Fig. 6.4. Of course, the same applies to any surface-parallel polarisation $\boldsymbol{E}_{\parallel}$. The generalisation of the above result,

$$\hat{E}_{\parallel R} + \hat{E}_{\parallel I} = 0 \quad \text{but} \quad \hat{B}_{\parallel R} - \hat{B}_{\parallel I} = 0, \tag{6.21}$$

will be needed later. Regardless of the sign, the fact remains that the values for the amplitudes of the incoming and outgoing wave are identical.

Refraction of light is a consequence of velocities being different

The well-known phenomenon called refraction is sketched in ◘ Fig. 6.5. Its only cause is that the speed of light differs in different materials according to ◘ Eq. (6.8). For an intuitive understanding, consider two points of the same

◘ **Fig. 6.4** The amplitudes of the electromagnetic fields which, at normal incidence, $\boldsymbol{k}_I = (0, -k, 0)$. In the plane of the surface, $\boldsymbol{E}$ is reversed, $\boldsymbol{B}$ is not

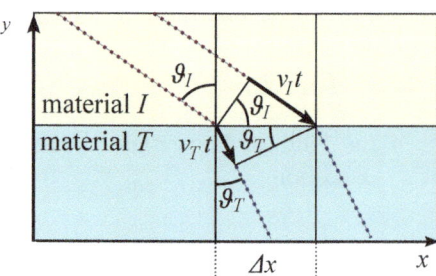

□ **Fig. 6.5** On the geometry of the refraction of a electromagnetic wave at a material interface. The arrows represent the distances covered in time t. The angles ϑ_I and ϑ_T for Surface normals can be found in two triangles, whose common hypotenuse is the distance Δx

6

wavefront. During a specific time interval t, a greater distance is travelled in the medium with the higher velocity. Let ϑ be the angle between the direction of wave movement and the surface normal then, as shown in □ Fig. 6.5, the equations

$$\Delta x \sin(\vartheta_I) = v_I t \quad \text{and}$$
$$\Delta x \sin(\vartheta_T) = v_T t$$

are valid. The ratio

$$\frac{\sin(\vartheta_I)}{\sin(\vartheta_T)} = \frac{v_I}{v_T} = \frac{n_T}{n_I} \tag{6.22}$$

with the refractive index $n = c/v$ is known as the *law of refraction*.

If $\sin(\vartheta_T)v_I/v_T > 1$, Eq. (6.22) can no longer be fulfilled. Then, the entire wave is reflected. The angle ϑ_T from which this occurs for a given material combination is the *critical angle*.

The proportion of refracted waves depends on the polarisation

While total reflection is limited to certain angular ranges, in general, waves are reflected and refracted simultaneously, as shown in □ Fig. 6.6. The proportion that is reflected can be obtained in a similar manner as the refraction laws for field lines (3.26, 3.27). The following lines will summarise the derivation.

The directions e_k of the wave vectors are shown in □ Fig. 6.6 and given by

$$e_{kI} = (\sin\vartheta_I, -\cos\vartheta_I, 0) \quad e_{kR} = (\sin\vartheta_I, \cos\vartheta_I, 0) \quad e_{kT} = (\sin\vartheta_T, -\cos\vartheta_T, 0).$$

The field components which are vertical to the surface (E_y and B_y) and those that are parallel (E_x, B_x, E_z and B_z) show different behaviours. If the interface between the substances "I" and "R" is placed the xz plane (see also ◘ Figs. 3.8 and 3.9) then, the following relations for the components lying in the plane of the interface appear as a result of Maxwell's equations (3.1):

$$\text{law} \rightarrow \qquad\qquad \text{close to the interface}$$

1. $\oint \mathbf{E}d\mathbf{l} = 0 \rightarrow \qquad E_{xI} + E_{xR} = E_{xT}$

2. $\oint \mathbf{E}d\mathbf{l} = 0 \rightarrow \qquad E_{zI} + E_{zR} = E_{zT}$

$$\text{(6.23)}$$

3. $\oint \mu^{-1} \mathbf{B}d\mathbf{l} = 0 \rightarrow \quad \mu_I^{-1} B_{xI} - \mu_I^{-1} B_{xR} = \mu_T^{-1} B_{xT}$

4. $\oint \mu^{-1} \mathbf{B}d\mathbf{l} = 0 \rightarrow \quad \mu_I^{-1} B_{zI} - \mu_I^{-1} B_{zR} = \mu_T^{-1} B_{zT}$

Here, the index I in $\mathbf{E}_I$ and $\mathbf{B}_I$ indicates the material passed by the incident wave.

Next, a distinction between *vertical polarisation* and *parallel polarisation* will be required. Here, the terms vertical and parallel define the orientation of the electric field relative to the interface (in ◘ Fig. 6.6, this is the xy plane).

1. **Vertical polarisation**: Here, the electric field vector is perpendicular to the xy plane, so it has the form $\mathbf{E} = (0, 0, E)$. According to Eq. (6.23.2), the field contributions have to obey

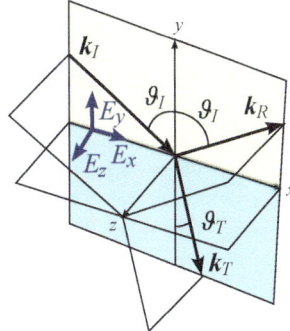

◘ **Fig. 6.6** An electromagnetic wave with the wave vector $\mathbf{k}_I$ hitting the interface between two substances. A fraction "R" is reflected, another, "T" refracted. The lengths of the vectors denoted by $\mathbf{k}$ are not to scale

$E_I + E_R = E_T$.

Using, $\boldsymbol{B} = \frac{1}{v}\boldsymbol{e}_k \times \boldsymbol{E}$, the magnetic fields are determined to be

$$\boldsymbol{B}_I = \frac{E_I}{v_I}(-\cos\vartheta_I, -\sin\vartheta_I, 0) \ ,$$

$$\boldsymbol{B}_R = \frac{E_R}{v_I}(\cos\vartheta_I, -\sin\vartheta_I, 0) \ \text{ and}$$

$$\boldsymbol{B}_T = \frac{E_T}{v_T}(-\cos\vartheta_T, -\sin\vartheta_T, 0) \ .$$

The x-component parallel to the surface can be written as

$$-\frac{E_I}{\mu_I v_I}\cos\vartheta_I + \frac{E_R}{\mu_I v_I}\cos\vartheta_I = -\frac{E_T}{\mu_T v_T}\cos\vartheta_T \ ,$$

as a consequence of ◻ Eq. (6.23.3). This allows either E_R or E_T to be eliminated. One can, for example, determine the field strength of the reflected wave as

$$\frac{E_R}{E_I} = \frac{\sqrt{\frac{\varepsilon_I}{\mu_I}}\cos\vartheta_I - \sqrt{\frac{\varepsilon_T}{\mu_T}}\cos\vartheta_T}{\sqrt{\frac{\varepsilon_I}{\mu_I}}\cos\vartheta_I + \sqrt{\frac{\varepsilon_T}{\mu_T}}\cos\vartheta_T} \ , \tag{6.24}$$

with the velocity $v = 1/\sqrt{\varepsilon\mu}$.

2. **Parallel Polarization**: Here, the electric field vector is in the xy plane so that the magnetic field $\boldsymbol{B} = (0, 0, B)$ is parallel to the surface. According to ◻ Eq. (6.23.4) one gets

$$\sqrt{\frac{\varepsilon_I}{\mu_I}}(E_I - E_R) = \sqrt{\frac{\varepsilon_T}{\mu_T}}E_T \ ,$$

with $E = vB$, and also according to ◻ Eq. (6.23.1)

$$(E_I + E_R)\cos\vartheta_I = E_T\cos\vartheta_T \ .$$

The result for the fraction of the reflected field is then

$$\frac{E_R}{E_I} = \frac{\sqrt{\frac{\mu_T}{\varepsilon_T}}\cos\vartheta_T - \sqrt{\frac{\mu_I}{\varepsilon_I}}\cos\vartheta_I}{\sqrt{\frac{\mu_I}{\varepsilon_I}}\cos\vartheta_I + \sqrt{\frac{\mu_T}{\varepsilon_T}}\cos\vartheta_T} \ , \tag{6.25}$$

i.e. an equation of a similar structure as the equation for transverse polarisation but with different signs and different factors in front of the cosine functions.

With the help of the refraction law ◻ (Eq. 6.22), the angle of the refracted wave can be eliminated using

$$\cos^2(\vartheta_T) = 1 - \frac{n_I^2}{n_T^2}\sin^2(\vartheta_I) \quad \text{(both polarisations)}. \tag{6.26}$$

As a result, the fraction of the reflected wave can be determined directly from the angle of incidence ϑ_I, provided the material properties are known.

Measurement of the Refractive Index

A popular method for determining the refractive index n is based on the fact that the proportion of reflected parallel polarised light is zero at a certain angle, the *Brewster angle* $\vartheta_{\text{Brewster}}$, as can be seen in ◻ Fig. 6.7. The angle depends on the refractive index.

Glass and air, like many other substances, have a negligible amount of magnetizability so that

$$\mu_r \approx 1 \rightarrow n \approx \sqrt{\varepsilon_r}$$

is a good approximation. Therefore, all factors of $\sqrt{\varepsilon/\mu}$ in ◻ Eqs. (6.24, 6.25) can be replaced by the refractive index n. For a given combination of refractive indices, the proportion of reflected light depends on the incident angle only (the law of refraction predicts the second). In ◻ Fig. 6.7 this is shown for $n_{\text{air}} = 1$ and $n_{\text{glass}} = 1.5$. At an angle of incidence of $\vartheta_{\text{glass}} = \arctan(n_{\text{air}}/n_{\text{glass}}) = 33.7°$ no parallel polarised light is reflected. According to the law of refraction, this corresponds to an incident angle

$$\vartheta_{\text{air}} = \arctan(n_{\text{glass}}/n_{\text{air}}) = \vartheta_{\text{Brewster}} = 56.3° .$$

The refractive index can be measured by varying the incident angle of LASER light ϑ_{air} on a glass surface. The reflected light is directed through a polarisation filter onto a photodiode. If "no" (actually, a minimum is determined), parallel polarised light is received, and the Brewster angle is found. This angle determines the refractive index.

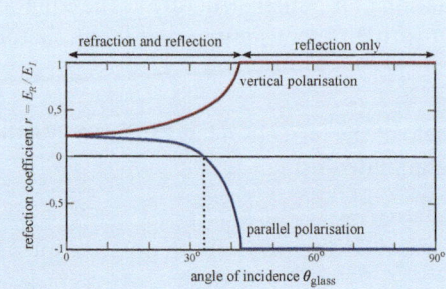

⬛ **Fig. 6.7** Reflection coefficient $r = E_R/E_I$ at the surface of glass ($n \approx 1.5$) in air as a function of the angle ϑ_{Glass} to the surface normal in the glas

6

6.3 **Conservation of Energy and Poynting's Theorem**

Conservation of energy in electrodynamic systems forms the basis of *Poynting's theorem*. It states that a system's timely loss of field energy equals the power of the radiated electromagnetic fields. This will be shown below. Following Griffith [2], the following strategy is pursued: First, the energy change dW and power $P = dW/dt$ due to the movement of the charge carriers within the fields are calculated. Then, the charge is eliminated using Maxwell's equations. This leads to an equation that relates the change in the system's field energy to the radiation power.[4]

The first step leads from the electrodynamic force acting on a charged particle that moves with the speed v to the power loss P:

$$\boldsymbol{F} = Q(\boldsymbol{E} + \boldsymbol{v} \times \boldsymbol{B}) \qquad \text{electrodynamic force}$$
$$\rightarrow dW = \boldsymbol{F}d\boldsymbol{r} = Q(\boldsymbol{E} + \boldsymbol{v} \times \boldsymbol{B}) \cdot d\boldsymbol{r} \quad \text{energy loss with } \boldsymbol{v} = d\boldsymbol{r}/dt$$
$$\rightarrow P = dW/dt = Q\boldsymbol{E} \cdot \boldsymbol{v} + \boldsymbol{v} \times \boldsymbol{B} \cdot \boldsymbol{v} \quad \text{power}$$

Because of $\boldsymbol{v} \times \boldsymbol{B} \cdot \boldsymbol{v} = 0$, the temporal energy loss (power) for a continuous charge distribution is then

$$P = \int \rho \boldsymbol{v} \cdot \boldsymbol{E} dV' = \int \boldsymbol{J} \cdot \boldsymbol{E} dV'. \tag{6.27}$$

---

4   In this original form of Poynting's theorem, no interactions of fields and matter are accounted for.

The disappearance of the B term is a consequence of the fact that a charge carrier passing through a magnetic field in free flight changes its direction but not its velocity

The second step is to find an expression for $J \cdot E$ that contains fields only. This is done using the Amere-Maxwell law (3.1.4). Multiplying the law by E gives an expression

$$
\begin{aligned}
J &= \nabla \times (\mu_0^{-1} B) - \tfrac{\partial}{\partial t}(\varepsilon_0 E) \\
\rightarrow E \cdot J &= E \cdot [\nabla \times (\mu_0^{-1} B)] - E \cdot \tfrac{\partial}{\partial t}(\varepsilon_0 E) \\
&= E \cdot [\nabla \times (\mu_0^{-1} B)] - \tfrac{1}{2}\tfrac{\partial}{\partial t}(\varepsilon_0 E^2),
\end{aligned}
$$

which already contains one term which can be interpreted as a loss of electric field energy: $[\partial(\varepsilon_0 E^2)/\partial t]/2$. Now, in the remaining expression $E \cdot [\nabla \times (\mu_0^{-1} B)]$ the loss of magnetic field energy can be sought. When seeking a $B \cdot \partial B/\partial t$ term, a connection between E and $\partial B/\partial t$, i.e. the Faraday-Henry law

$$
\nabla \times E = -\frac{\partial B}{\partial t}
$$

is the obvious choice. And using the product rule

$$
\nabla \cdot (E \times B) = B \cdot (\nabla \times E) - E \cdot (\nabla \times B)
$$

the connection is found: The result

$$
E \cdot J = -\left(\frac{1}{2\mu_0}\frac{\partial B^2}{\partial t} + \frac{\varepsilon_0}{2}\frac{\partial E^2}{\partial t} \right) - \frac{1}{\mu_0}\nabla \cdot (E \times B) \quad \text{(without matter)}
$$

(6.28)

contains a term

$$
S = \frac{1}{\mu_0}E \times B \quad \text{(without matter)},
$$

(6.29)

which is called *Poynting vector*. At first glance, this looks asymmetrical and may be wrong. Why is there a Factor $1/\mu_0$, but not a ε_0? The reason becomes clear if S is written as an *energy flux density*, i.e., the product of a velocity and an energy density. With no matter involved, the velocity should be the vacuum speed of light $c = 1/\sqrt{\varepsilon_0\mu_0}$ (see ◼ Eq. (6.8) below). Then, the equation

$$S = c \left(\sqrt{\varepsilon_0} E \right) \times \left(B / \sqrt{\mu_0} \right)$$

appears that is easier to interpret. The Poynting vector is proportional to the harmonic mean of the energy densities of the fields E and B. In addition, the field constants appear in a balanced manner. And there is evidence that the Poynting vector is a radiation term. If the electric field contains as much energy as the magnetic field, and if E and B are perpendicular to each other, then the magnitude of the Poynting vector is equal to the product of the total, i.e. electric plus magnetic field energy density and the speed of light.

With the help of Gauss' theorem, the power (6.27) can be written down explicitly

$$P = \int_V J \cdot E \, dV' = -\int_V \left(\frac{1}{2\mu_0} \frac{\partial B^2}{\partial t} + \frac{\varepsilon_0}{2} \frac{\partial E^2}{\partial t} \right) dV' - \int_{\partial V} S \cdot da' \,.$$

The environment changes, the Poynting vector does not

Energy conservation applied to charges traversing fields led to Poynting's theorem (6.28). However, combining Ampere-Maxwell's law with the Faraday and Henri law suffices to derive Poynting's theorem. This fact can be used to derive power balances for fields in matter (3.7). Then, the same reasoning as used for the derivation of ◻ Eq. (6.28) leads to

$$E \cdot J_{\text{free}} = -\left[(\mu^{-1} B) \cdot \frac{\partial B}{\partial t} + E \cdot \frac{\partial (\varepsilon E)}{\partial t} \right] - \nabla \cdot \left[E \times (\mu^{-1} B) \right] \quad \text{(anisotropic material)}$$

$$E \cdot J_{\text{free}} = -\left(\frac{1}{2\mu} \frac{\partial B^2}{\partial t} + \frac{\varepsilon}{2} \frac{\partial E^2}{\partial t} \right) - \nabla \cdot (E \times \frac{B}{\mu}) \quad \text{(isotropic material)},$$

$$\text{(6.30)}$$

where for anisotropic materials, i.e. materials rotating fields, ε and μ^{-1} can be read as matrices. In any case $\varepsilon = \varepsilon_0 \varepsilon_r$ and $\mu^{-1} = \mu_0^{-1} \mu_r^{-1}$ may be arbitrary functions of the field strengths and locations.

A simple interpretation of ◻ Eq. (6.30) appears, if ε_r and μ_r^{-1} are constant factors: They indicate by which factor the energy density is increased by the presence of matter (usually due to its polarisation). And they modify the velocity term

$$\text{without matter} \rightarrow \text{inside of matter}$$
$$c \qquad \rightarrow c / \sqrt{\varepsilon_r \mu_r}$$

with the result of a speed reduction. The Poynting vectors in ◻ Eq. (6.30) look identical. So, one may conclude that the formula

$$S = \left[E \times (\mu^{-1} B)\right] \quad \text{(always)},\tag{6.31}$$

which describes the power loss density through fields is valid under all conditions. The results of its application are often counter-intuitive, as the following example shows:

Cables transport electricity, but not energy

With the help of Poynting's theorem, it can be shown that the energy transport of electrical supply lines does not take place within these lines, but in the fields between them. First, it will be demonstrated that the Poynting vector within a wire has no longitudinal component.

Let a wire of radius R and length l be exposed to a voltage U in the longitudinal direction. Then, a current I flows and creates a magnetic field within the wire. Its strength increases with the radius and reaches $|B| = \mu I/(2\pi R)$ at the surfaces of the wire. The field lines are, as shown in ◻ Fig. 6.8, parallel to the wire surface. And there is an electric field $|E| = U/l$ along the wire, also parallel to the surface but perpendicular to the magnetic field. The magnitude of the Poynting vector and the energy loss are then given by

$$|S| = \frac{1}{\mu} \frac{U}{l} \frac{\mu I}{2\pi R} \rightarrow P = UI$$

because the surface of a piece of wire is $2\pi Rl$. And the vector S is pointing out of the wire. This means that the energy flows out of the wire and the power is exactly as large as the power loss due to the ohmic resistance.[5] So, the Poynting vector describes the power density of thermal radiation.

◻ **Fig. 6.8** The electromagnetic fields E and B as well as the Poynting vector S in a current-carrying wire. The Poynting vector points to outside

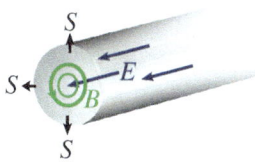

---

5   To be precise: This result is only true when the wire has reached its final temperature so that the wire itself absorbs no energy.

🔲 Figure 6.9 shows the fields between two wires carrying electricity back and forth. For reasons of symmetry, straight, parallel wires must have electrical and magnetic field lines lying in a plane perpendicular to the wires. So, if there is a socket mounted on a wall, the lines could be drawn onto this wall. The Poynting vector must point in a direction parallel to the wires. The energy transport is tied to the wires but takes place outside of them.

Energy transport comes with the transport of momentum

Electromagnetic fields transporting energy always transport a momentum P as well. Starting with $dP/dt = Q(E + v \times B)$, this can be shown in a complex calculation (see e.g. [2]) using Maxwell's equations.

Richard P. Feynman [3] suggested an alternative reasoning: According to a fundamental theorem of relativistic mechanics, every energy transfer also involves a momentum transfer. If the energy flux density (energy per time interval and area, i.e. power density) is divided by the square of the speed of light, the result is the momentum density (momentum per volume). If this momentum density is called g it must consequently be given by

$$g = \frac{1}{c^2} S \qquad\qquad (6.32)$$

for electromagnetic fields. The formula looks simple, but its interpretation is not: While the Poynting vector represents the energy flux density, i.e. it contains the velocity of energy transport, $|g|$ is the momentum density within a volume. The momentum flux density is only obtained when momentum density is multiplied by the velocity of the the electromagnetic field. As with any flux density, its product is with an area a is the momentum flow rate through

🔲 **Fig. 6.9** Simulation of the magnetic field in a live socket. The field is strongest between the lines. The electric field lines (not shown) connect the wires. They are always perpendicular to the magnetic field lines. The Poynting vector points out of the socket

this area. The change of momentum at a surface exposed to electromagnetic radiation is consequently given by

$$\frac{\mathrm{d}p}{\mathrm{d}t} = c(g \cdot a),$$

where c is the velocity vector of the radiation. If all radiation is absorbed on the surface, $F = \mathrm{d}p/\mathrm{d}t$ is the force acting on the surface.

Radiation Pressure

The best-known example of momentum transfer by electromagnetic waves is the radiation pressure on comets near the sun. Its amount can be derived from the power density of the solar radiation observed on earth $|S_{\text{earth}}| \approx 1\,\mathrm{kW/m^2}$. From ▪ Eq. (6.32) one gets

$$|S_{\text{sun}}| = |S_{\text{earth}}| \left(\frac{R_{\text{earth orbit}}}{R_{\text{sun}}} \right)^2$$
$$F = a_{\text{comet}} \, c |g_{\text{sun}}|$$

for a comet close to the sun's surface. The factor c behind the comet cross-sectional area a_{comet} turns the momentum density into a momentum flux density. For example, a comet having a cross-section $a_{\text{comet}} = 10\,\mathrm{m^2}$ is exposed to a force up to $F = 1.54\,\mathrm{N}$.

6.4 Simplifications for Slowly Changing Fields

Calculating quasi-statically means calculating with less effort

Quasi-static calculations are those in which systems with changing fields are calculated using the same methods as static systems. They are prevalent in alternating current technology. The formal set-up is summarised in ▪ Table 6.1. It assumes that the quantities shown there do not depend on time apart from the factor $e^{\mathrm{j}\omega t}$, and that they can be handled like time-independent quantities in every other respect. The extensions into the complex plane are to simplify calculations. The measurable quantities are then the imaginary parts $u = \Im(\underline{u}) = \hat{U}\sin(\omega t + \phi_U)$ etc.. The harmonious progression over time implies that the systems regularly go through the same states. Therefore they become in the literature often as *quasi stationary*.

■ **Table 6.1**   Replacement rules of classical alternating current technology

DC quantity	I	U	E	B
AC quantity	$\underline{i} = \underline{I}e^{j\omega t}$	$\underline{u} = \underline{U}e^{j\omega t}$	$\underline{E} = \underline{\hat{E}}e^{j\omega t}$	$\underline{B} = \underline{\hat{B}}e^{j\omega t}$
Amplitude	$\underline{I} = \hat{I}e^{j\phi_I}$	$\underline{U} = \hat{U}e^{j\phi_U}$	$\underline{\hat{E}} = \hat{E}e^{j\phi_E}$	$\underline{\hat{B}} = \hat{B}e^{j\phi_B}$

Next, we will discuss which prerequisites make such calculations trustworthy. However, the following restriction applies: As long as the exact solutions are unknown, one cannot be 100% sure that the quasi-static approximation is reasonable. The basis of the approximations is, therefore, rather practical reason than exact science.

Maxwell's equations contain the criteria

Quasi-static calculations are justified if and only if the terms which only occur in the dynamic theory have negligible effects. A look at Maxwell's equations shows that the terms $\partial E/\partial t$ and $\partial B/\partial t$ make the difference. If they are neglectable, the laws of statics also apply to harmoniously oscillating quantities. Under these conditions, magnetic fields are created exclusively by currents. If the $\partial B/\partial t$ term can be neglected, the Faraday-Henry law is reduced to the effect of the Lorentz force, as was shown in ■ Eqs. (3.20) and ■ (3.21). And electric fields are exclusively generated by charges.

Both $\partial E/\partial t$ and $\partial B/\partial t$ have dimensions. But whether something is negligible can only be determined based on a dimensionless ratio (which usually must be much smaller than one). Therefore, the following is checked:

1. Can the energy of the magnetic field that is caused by the oscillation of the electric field be neglected because $W_B \ll W_E$?
2. Can the energy of the electric field that is caused by the oscillation of the magnetic field be neglected because $W_E \ll W_B$?
3. Can the power of electromagnetic radiation be neglected relative to the ohmic loss of power. When is $P_{\text{radiation}} \ll P_\Omega$ fulfilled?

For this purpose, applications are being sought in which the dynamic behaviour can be calculated analytically.

Capacitor behaviour can be calculated stationary (almost always)

The behaviour of a capacitor is given by $Q = CU$ due to the properties of the electric field. However, according to Ampere-Maxwell's law, every change in the field creates a magnetic field. For a capacitor with circular electrodes with radius r and a distance of d its strength (see $\blacksquare$ Problem 3.9) at a distance a from the centre of the circle is given by

$$B = \frac{1}{2}\alpha\mu\varepsilon\frac{\partial E}{\partial t} \; .$$

An alternating voltage $E = \hat{E}\sin(\omega t)$ therefore creates a magnetic field $B = \frac{1}{2}\mu\varepsilon\omega\hat{E}\cos(\omega t)$. This contains an energy

$$\begin{aligned}
W_E &= (\tfrac{1}{2}\varepsilon E^2)\cdot(\pi r^2 d)\\
&= \tfrac{1}{2}\varepsilon\pi r^2 d\hat{E}^2\sin^2(\omega t) \; .
\end{aligned}$$

In contrast, the energy of the magnetic field is

$$\begin{aligned}
W_B &= \int_V \frac{B^2}{2\mu}\,\mathrm{d}V'\\
&= \frac{1}{2\mu}2\pi d\int_0^r \left(\frac{\alpha'}{2}\mu\varepsilon\omega\hat{E}\cos(\omega t)\right)^2 \alpha'\,\mathrm{d}\alpha'\\
&= \tfrac{\pi}{16}\mu\varepsilon^2\hat{E}^2\cos^2(\omega t)\mathrm{d}r^4\omega^2 \; ,
\end{aligned}$$

so that, on average, the ratio of the energies is

$$\frac{W_B}{W_E} = \frac{1}{8}\mu\varepsilon r^2\omega^2 = \frac{1}{8}\left(\frac{r\omega}{v}\right)^2$$

with $v = 1/\sqrt{\varepsilon\mu}$ being the velocity. Consequently, the behaviour of a circular capacitor can be calculated quasi-statically if the inequality

$$\frac{1}{8}\left(\frac{r\omega}{v}\right)^2 \ll 1 \tag{6.33}$$

is fulfilled.

A numerical example: For a capacitor with $C = 1\,\mu\text{F}$ being the value for the capacitance having a gap between the electrodes $d = 1\,\mu\text{m}$, the maximum energy content of the magnetic field is just as large as that of the electric field at a frequency $f_g = 711\,\text{MHz}$. If it is taken into account that real capacitors are wound or folded, i.e. radii as large as with plate capacitors never occur, one can be confident that the magnetic field inside a capacitor is hardly ever relevant in technical applications.

Coils contain electric fields

The magnetic field of a toroidal coil can be calculated analytically. As shown in ◘ Problem 6.5, the energy of the electric field due to the oscillation of the magnetic field is negligible if

$$\frac{W_E}{W_B} = \frac{1}{12}\left(\frac{\omega r}{c}\right)^2 \ll 1\,, \tag{6.34}$$

where r is the larger of the two radii of a torus, is satisfied. For $v \approx c$, the energy ratio is almost exactly the same the ratio of the field energies in the capacitor, if E and B are swapped.

While toroidal coils carry the entire magnetic field within themselves, coils with open ends do not. They emit electromagnetic waves. If the frequencies are high enough, they are used in mass products like RFID [4] used as transmitting antennas.

Small dipoles radiate small amounts of energy

The power of radiating dipoles $P_{\text{radiation}}$, grows in proportion to the square of the current $i = \hat{I}\sin(\omega t)$-just as the power loss of an ohmic resistance.

By analogy, the proportionality factor *radiation resistance* $R_{\text{radiation}}$ is introduced so that the power may be written as

$$P_{\text{radiation}} = \frac{1}{2}\hat{I}^2 R_{\text{radiation}}\,. \tag{6.35}$$

For dipoles whose length L is significantly smaller than the wavelength of the radiated wave is ($L \ll \lambda$), the radiation resistance can be found analytically [2] to be

$$R_{\text{radiation}} = \frac{\pi}{6}\sqrt{\frac{\mu_0}{\varepsilon_0}}\left(\frac{L}{\lambda}\right)^2 \approx 197\,\Omega\left(\frac{L}{\lambda}\right)^2 \quad \text{(short dipole)}. \tag{6.36}$$

The radiation power, therefore, sharply increases with the length L for $L \ll \lambda$. The closer the length L approaches half of the wavelength, the less precise ◘ Eq. (6.36) becomes. For the most popular dipole antenna, the "$L = \lambda/2$ antenna", a numerical calculation gives $R_{\text{radiation}} \approx 73\ \Omega$, which is slightly more than expected from the short dipole approximation.

The constant ratio of length and wavelength corresponds a fixed product of the angular frequency ω and the length. Therefore, the condition for a dipole antenna can be written as

$$L^2 \ll \left(\frac{\lambda}{2}\right)^2 \quad \rightarrow \quad \left(\frac{\omega L}{\pi c}\right)^2 \ll 1. \tag{6.37}$$

In ◘ Fig. 6.10, the limit $L = \lambda/2$ is drawn as a solid line. The radiation power is relevant to the extent that it approaches the ohmic power $R\hat{I}^2/2$. In ◘ Fig. 6.10, the dashed line derived from ◘ Eq. (6.36) shows the condition

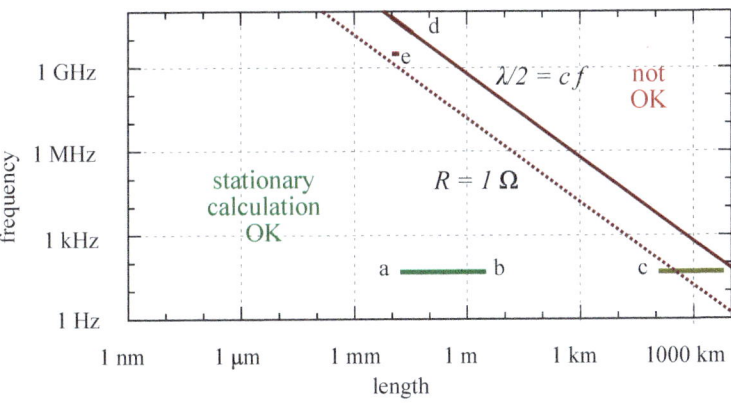

a-b home equipment, power stations
c pan american 50 Hz supply nets
d 5G antenna
e Intel I7 processor

◘ **Fig. 6.10** Limits for using stationary formulas set by electric dipole radiation. The maximum frequency f is shown as a function of the length L. Solid line: $\lambda/2$–dipole, dashed line: radiant power equals an ohmic power of $R = 1\ \Omega$

$$f = \frac{c}{L}\sqrt{\frac{6R\sqrt{\varepsilon_0}}{\pi\sqrt{\mu_0}}} \approx \frac{c}{L}\sqrt{\frac{R}{197\Omega}} \tag{6.38}$$

for the value $R = 1\ \Omega$. At this length and frequency, the power loss of an ohmic resistor whose value is $1\ \Omega$ equals the radiated power. The lower this resistance, the lower the frequencies accompanied by considerable radiation. This condition can also be expressed in a similar way to ● Eq. (6.37):

$$\left(\frac{\omega L}{\pi c}\right)^2 \ll \sqrt{\frac{R}{12,3\ \Omega}}. \tag{6.39}$$

Its meaning is quite different. Here, the power balance of the entire system is considered, not only the fields.

An example from electronics: Radiation is not to be neglected in microprocessors. For example, the Intel I7 processor [5] has a supply voltage of $V_{DD} = 1.52$ V. Depending on the variant, it takes a maximum currents of $35–85$ A, to ohmic resistances in the 30 mΩ range.

Commonalities strengthen the confidence in approximations

Except for a few per cent, the capacitor, the coil and the dipole antenna lead to the same condition

$$\left(\frac{\omega L}{\pi c}\right)^2 \ll 1,$$

where L is the length of the system. Under this boundary condition, it seems inevitable that the fields generated by fields are negligible relative to those generated by currents and charges.

The dipole antenna teaches us that radiation adds significantly to the power loss when large currents accompany small voltages. The maximum frequency for the the usability of quasi-static approximations therefore decreases below the above limit if the ohmic resistance drops below $R = 12\ \Omega$.

6.5 Stationary Transmission Line Theory

Transmission lines are as indispensable in electrical engineering as veins are in biology. The classical theory, expressed by the *telegraph equations*, is based on the assumption that stationary calculations are sufficiently accurate. The starting point is the model shown in ◘ Fig. 6.11. It represents a small fraction of the entire line.

The book [6] shows how the transmission line equations can be derived from the model shown in ◘ Fig. 6.11. The procedure is as follows: first, the line is divided into N parts. Then, each part is modelled as shown in ◘ Fig. 6.11, and finally, the behaviour of the chain connection of all N elements is calculated. In the limit, $N \to \infty$, one gets the behaviour of a line with all impedances continuously distributed over the length.

For a twin cable line with an ohmic resistance R and an inductance L along the entire path back and forth and a capacitance C between the individual lines one gets

$$\underline{u}_1 = \cosh(\underline{g})\,\underline{u}_2 + \underline{Z}_0 \sinh(\underline{g})\,\underline{i}_2$$
$$\underline{i}_1 = \frac{1}{\underline{Z}_0}\sinh(\underline{g})\,\underline{u}_2 + \cosh(\underline{g})\,\underline{i}_2 \qquad (6.40)$$

with the *transmittance g* and the *characteristic impedance* $\underline{Z}_0$ (also called *wave resistance*) given by

$$\underline{g}(\text{line}) = \sqrt{j\omega RC - \omega^2 LC}$$
$$\underline{Z}_0(\text{line}) = \sqrt{\frac{L}{C} - \frac{jR}{\omega C}} . \qquad (6.41)$$

These equations are referred to as *telegraph equations*. If $\underline{g} = 0$, current and voltage are the same at both ends of the line. Therefore, the line's characteristics are determined only by two complex parameters.

According to ◘ Eq. (6.41), the ohmic resistance has no significant influence on the characteristics if the condition

$$\omega L \gg R \quad (\text{then } R \text{ is negligible})$$

is fulfilled. Surprisingly, this does not depend on the capacitance of the line.

◘ **Fig. 6.11** Model of a small piece of a transmission line with complex currents and voltages at input and output

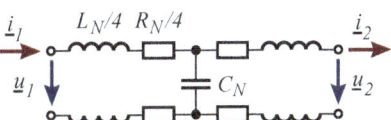

In the following paragraphs, the relation of the two parameters g and $\underline{Z}_0$ and the fields accompanying the lines shall be examined. After that, the consequences for the line's conduction characteristics shall be examined.

Ideal cables have no losses and negligible fields in the conductors

Generally, one has to distinguish between *lossless* ($R = 0$) lines and *ideal* ones ($R = 0$ *and* Φ_B negligible within the conductors). First, the ideal coaxial cable shall be discussed. Its capacitance follows from ◘ Eq. (4.36). The inductance is given in the solution of ◘ Problem 6.6. If the radii delimiting the dielectric are called r_i and r_a,

$$C_{\text{coax}} = \frac{2\pi\varepsilon l}{\ln(r_a/r_i)}$$

$$L_{\text{coax}} = \frac{\mu l}{2\pi} \cdot \ln(r_a/r_i)$$

are the capacitance and the inductance of the ideal coax cable. These equations contain surprisingly similar terms. Therefore, the following short expressions give the line parameters:

$$\underline{g}_{\text{coax}} = j\omega l\sqrt{\mu\varepsilon} \quad \text{and} \quad \underline{Z}_{0,\text{coax}} = \frac{1}{2\pi}\ln\left(\frac{r_a}{r_i}\right)\sqrt{\frac{\mu}{\varepsilon}} \tag{6.42}$$

The transmittance g turns out to be completely independent of the radii. The equations for the ideal twin line look quite similar

$$C_{\text{twin line}} = \frac{\pi\varepsilon l}{\ln\left(\frac{d-r}{r}\right)} \quad \text{and}$$

$$L_{\text{twin line}} = \frac{\mu l}{\pi} \cdot \ln\left(\frac{d-r}{r}\right)$$

and lead to

$$\underline{g}_{\text{twin line}} = j\omega l\sqrt{\mu\varepsilon} \quad \text{and} \quad \underline{Z}_{0,\text{twin line}} = \frac{1}{\pi}\ln\left(\frac{dr}{r}\right)\sqrt{\frac{\mu}{\varepsilon}}, \tag{6.43}$$

which shows:

> **Design Factors**
> The transmittance of an ideal cable depends exclusively on its length and the material between the forward and return lines. In contrast, its characteristic impedance is independent of the length. It is determined by the cable cross-section and the material.

This has consequences for engineering: The transmittance can only be influenced by the choice of the material, not by the geometry. In contrast, the wave resistance can be adjusted by choosing a certain cross-section. The logarithmic singularity sets the standard for manufacturing tolerances when producing small conductor radii.

Line termination makes the difference

According to ◻ Eq. (6.41), lossless cables have a purely imaginary transmittance $\underline{g} = j\omega\sqrt{LC}$ and a purely real characteristic resistance $\underline{Z}_0 = \sqrt{L/C}$. Their relations to the voltages and currents shall be examined next. It turns out that the termination of the line makes all the difference.

With $\cosh(j\alpha) = \cos(\alpha)$ and $\sinh(j\alpha) = j\sin(\alpha)$ the lossless line can be described by

$$\underline{u}_1 = \cos(\omega\sqrt{LC})\,\underline{u}_2 + j\underline{Z}_0\sin(\omega\sqrt{LC})\,\underline{i}_2$$

$$\underline{i}_1 = \frac{j}{\underline{Z}_0}\sin(\omega\sqrt{LC})\,\underline{u}_2 + \cos(\omega\sqrt{LC})\,\underline{i}_2 \,.$$

Now, three cases may be distinguished and analysed. What happens in the cases of a short circuit ($\underline{u}_2 = 0$), idle ($\underline{i}_2 = 0$) and *matching* ($\underline{u}_2 = \underline{Z}_0 \cdot \underline{i}_2$) when powered by an ideal voltage source?
- Idle: $\underline{u}_1 = \cos(\omega\sqrt{LC})\underline{u}_2$, $\rightarrow \underline{u}_2$ diverges at $\omega\sqrt{LC} = \frac{\pi}{2} \pm N \cdot \pi$.
- Short circuit: $\underline{i}_1 = \frac{1}{j\underline{Z}_0}\cot(\omega\sqrt{LC})\underline{u}_1$, $\rightarrow \underline{i}_1$ diverges at $\omega\sqrt{LC} = \pm N \cdot \pi$.
- Matching: $\underline{u}_1 = e^{j(\omega\sqrt{LC})}\underline{u}_2$ and $\underline{i}_1 = e^{j(\omega\sqrt{LC})}\underline{i}_2$, $\rightarrow$ only phase shift, all amplitudes remain unchanged.

This compilation demonstrates the importance of the characteristic impedance. If a load impedance equals the characteristic impedance of its connecting cable, then, apart from a phase shift, the load is supplied in the same way as if it were directly connected to the source. This case is referred to as *matching*.

Non-ideal sources are used for high frequencies to ensure that no disasters will happen, even if the line is either open or shorted. State-of-the-art is the use of both matching sources and matching loads.

Long lines may create technical problems

The transmission g is proportional to the cable length. Consequently, it sets the scale whether a line is *short* or *long*. According to ◾ Eqs. (6.42, 6.43)

$$l \ll \frac{1}{\omega\sqrt{\varepsilon\mu}}$$

means that "the ideal line is short". The factor $1/\sqrt{\varepsilon\mu}$ has already been identified as speed in a different context. For a lossless line, it has the same meaning as will be shown next.

A phase shift always stands for a time difference: From $\underline{u}_1 = e^{j(\omega\sqrt{LC})}\underline{u}_2$ one may deduce that

$$\underline{U}_2 e^{j\omega t} = e^{-j\omega\sqrt{LC}}\underline{U}_1 e^{j\omega t} = \underline{U}_1 e^{j\omega(t-\sqrt{LC})} \ .$$

The output voltage lags behind the input voltage by a time $\Delta t = \sqrt{LC}$. Next, the length Δl that belongs to this time difference must be determined. To do this, one uses the inductance per unit length $L_l = L/\Delta l$ and the capacitance per unit length $C_l = C/\Delta l$ so that

$$\Delta t = \sqrt{LC} = \sqrt{L_l C_l} \cdot \Delta l \ . \tag{6.44}$$

◾ Equation (6.44) allows the signal velocity $v = \Delta l/\Delta t$ to be written as

$$v = \frac{1}{\sqrt{L_l C_l}} \qquad \text{(lossless line)}$$

$$v = \frac{1}{\sqrt{L_l C_l}} = \frac{1}{\sqrt{\varepsilon\mu}} \quad \text{(ideal line)} \tag{6.45}$$

as shown by a comparison with ◾ Eqs. (6.42, 6.43).[6] Empirically, a long line has significantly different voltages in different places simultaneously.

---

[6] ◾ Equation (6.45) can be used to verify manufacturer information. If more than the speed of light comes out, scepticism is warranted.

Short lines have non-trivial properties, too

The technical challenges in the energy supply sector are entirely different from those in communications technology. This will be demonstrated using the example of an underground cable.

Underground Power Supply

A new city district shall be supplied with electricity using a three-phase underground cable. It has a length of $l = 5$ km. Each phase is connected to a voltage source delivering $U_{eff} = 110$ kV at $f = 50$ Hz and individually shielded. The conductos are made from aluminum ($\rho_{Al} = 2.64 \cdot 10^{-8}$ Ωm). VPE (cross-linked polyethylene, $\varepsilon_r = 2.2$) with a dielectric strength of $E_{max} = 40$ kV/mm is used for insulation. The cable resistance per unit length shall not exceed $R_{max}/l = 0.1$ Ω/km for each phase.

Geometry (without corrosion protection, etc.):
The radius of the inner conductor r_i is determined by the specific resistance of aluminum:

$$R = \rho_{Al}\frac{l}{\pi r_i^2} \rightarrow r_i = \sqrt{\frac{\rho_{Al}l}{\pi R_{max}}} \rightarrow r_i = 9.17 \text{ mm}$$

The insulator's outer radius r_a results from the material's dielectric strength. Its determination must take into account that the electrical field strength decreases from the inside to the outside, and the peak value $\hat{U}$ is greater than the effective value of the voltage.

$$E(r) = \frac{U}{r\ln(r_a/r_i)} \rightarrow r_a = r_i \, e^{\frac{\sqrt{2}\,U_{eff}}{E_{max}\,r_i}} \rightarrow r_a = 14.0 \text{ mm}$$

Electrical Properties:
In this case, the condition $l \ll 1/(\omega\sqrt{\varepsilon\mu})$ leads to $l \ll 644$ km. Therefore, the 5 km long cable can be viewed electrically *short*: The value for the voltage is almost the same everywhere in the cable at any point in time.

According to the solution to ◻ Problem 4.11, the parasitic capacitance of each phase is

$$C = \frac{2\pi \varepsilon l}{\ln(r_a/r_i)} \rightarrow C = 1.45\,\mu\text{F}.$$

This parasitic capacitor must be charged and discharged 100 times per second. The reactive charging currents lead to ohmic losses: In a piece of line of length Δx at a distance x from the beginning of the line, the loss is $\Delta P = \rho_{\text{Al}} \Delta x\, I_{\text{eff}}^2(x)/(\pi r_i^2)$ The current decreases linearly with distance. Therefore, the power loss in the inner conductor is

$$P_{\text{inner}} = \int_0^l \frac{\rho_{\text{Al}}}{\pi r_i^2} I_{\text{eff},0}^2 \cdot \left(1 - \frac{x'}{l}\right)^2 dx' = \frac{1}{3} R_{\text{inner}} I_{\text{eff},0}^2,$$

where R_{inner} is the total resistance of the line (here: 0.5Ω) and $I_{\text{eff},0}$ is the effective value for the current at the beginning of the line. The latter is the one which must supply the entire parasitic capacitance: $I_{\text{eff},0}^2 = (\omega\, C\, U_{\text{eff}})^2$. The power loss calculation must also include the (significantly larger) total resistance through the shielding, R_{shield}. The result

$$P = \frac{1}{3}(R_{\text{inside}} + R_{\text{shield}}) \cdot (\omega\, C\, U_{\text{eff}})^2$$

shows that these losses increase sharply with voltage. And it formulates a maximum condition for the resistance of the shielding for a given power loss when idling.

Another maximum condition for the resistance of the shield is provided by its voltage fluctuations caused by the capacitive coupling. These increase with the distance to the beginning of the line according to

$$U(x) = \frac{\rho_{\text{shield}}}{A_{\text{shield}}} \int_0^x I(x')dx' \rightarrow U_{\text{eff}}(x) = \frac{R_{\text{shield}} I_{\text{eff},0}}{l} \left(x - \frac{x^2}{2l}\right)$$

and have a maximum at the end of the line the end of the line $U_{\text{eff}}(l) = I_{\text{eff},0} \cdot R_{\text{shield}}/2$ their maximum.

According to ▪ Problem 6.6, the inductance (ignoring the shielding) is

$$L = \frac{l\mu_0}{4\pi} \left[1 + 2\ln\left(\frac{r_a}{r_i}\right)\right] \rightarrow L = 923\,\mu\text{H}.$$

The signal speed can be determined from the inductance and the capacitance. The result

$$v = \frac{1}{\sqrt{L_l C_l}} \rightarrow v = 1.37 \cdot 10^8 \, \frac{\text{m}}{\text{s}}$$

is just below half the speed of light. The speed reduction is due to the factor ε_r occurring in the capacitance and due to the significantly increased inductance because of the large line cross-section compared to an ideal cable.

6.6 Concluding Remarks

May the end of this book be a beginning for you. You know the terms necessary to understand electrodynamics and recognise the basic relationships. You are, figuratively speaking, standing in front of a wide-open door. Behind this door is a wealth of knowledge, experience and know-how.

Treat these treasures with respect, but not uncritically. Because those who only follow authorities will always remain second-class. And many books and articles, even those of recent date, are coated with a patina from the ideas of former centuries. Einstein's guideline quoted in the preface leads to the following criterion: The greater the number of technical terms, the thicker the patina. Trust your own reasoning, and you will prevail. Samuel Ting put it this way: "If you do something well, you always find something new."

Last but not least, should it really happen that you don't understand something, this may be not your fault, but due to the book you are reading (except for this one, of course).

6.7 Problems

6.1 A coaxial cable (see also ◘ Fig. 5.18), whose dielectric reaches from the radius R_1 of the inner to the inner radius of the outer conductor R_2, is assumed to have negligible ohmic resistances of the conductors. At a voltage of $U = 10000$ V, a current of $I = 10$ A flows. Please calculate the Poynting vector and energy transport, i.e. the power through the dielectric.

How would the behaviour of the cable change if a Dielectric with $\varepsilon_r > 1$ was used?

6.2 Using the sun's radiation pressure to accelerate spaceships has been suggested. How large would a solar sail have to be near the Earth so that it is pushed away from the sun with a force of 1 N? The power density of solar radiation on Earth is $P/a \approx 1 \, \text{kW/m}^2$.

6.3 An electric field component depends on the location x and the time t according to

$$E_x = \hat{E}_x \cos(kx)\cos(kvt)$$

from [3]. Please check whether the field is a solution of the wave equation $\Delta E = \mu\varepsilon\partial^2 E/(\partial t)^2$. Is this a d'Alembert solution? And if so, please comment on the field quality!

6.4 An electromagnetic wave travels in a vacuum according to a $E = (0, \hat{E}, 0) \cdot \sin(\omega t - k_x x)$. Please determine its accompanying magnetic field B and show that $\hat{E} = c\hat{B}$ applies. Try to formulate the result independently of the choice of the coordinate system by expressing B as a function of E and the wave vector k.

6.5 A toroidal coil has N turns and the radii R and r shown in ▫ Fig. 6.12. Inside, a current of angular frequency ω creates a magnetic field, which in turn creates an electric field. Please determine the ratio of the electric field energy relative to the magnetic field energy, neglecting the magnetic field created by the electric field.

▫ **Fig. 6.12** Illustrating
▫ Problem 6.5: Coil wound as a torus and the definition their two radii r and R (Photo from de.wikipedia.org)

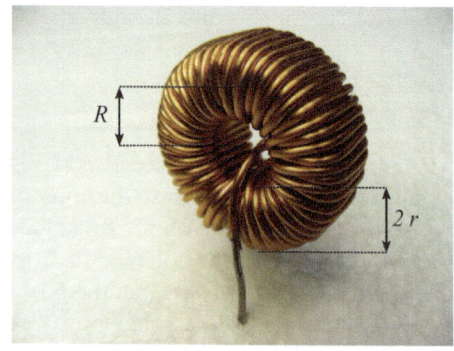

6.6 Determine the inductance per unit length of the arrangement shown in
◻ Fig. 5.18, starting from the known field components from the solution to
◻ Problem 5.9 The currents of the forward and return lines are the same.

6.7 Show that the telegraph equations 6.40 are symmetric under the swapping
of the line connections ($\underline{u}_1 \leftrightarrow \underline{u}_2$ and $\underline{i}_1 \leftrightarrow -\underline{i}_2$).

6.8 Usually, the fields accompanying an ideal line move significantly slower
than the speed of Light. The dominant factor is $1/\sqrt{\varepsilon_r}$ of the dielectric.
Shouldn't the power transmitted by the fields also decrease accordingly? And
if not-why not?

6.8 Solutions

6.1 According to Ampere's law (see ◻ Exercise 5.9), the strength of the magnetic field for radii r in the range $R_1 < r < R_2$ is

$$B = \frac{\mu I}{2\pi r} \, ,$$

and the field lines form concentric circles. The charge Q is evenly distributed
over the length l to determine the electric field. Using Gauss' theorem, field
strength and voltage turn out to be

$$E = \frac{Q}{2\pi \varepsilon r l} \quad \text{and} \quad U = - \int_{R_1}^{R_2} E dl = \frac{Q \ln(R_2/R_1)}{2\pi \varepsilon l} \, ,$$

and therefore

$$E = \frac{U}{r \ln(R_2/R_1)} \, .$$

The field lines point from the inner conductor to the outer conductor and are,
therefore, perpendicular to those of the magnetic field. The vector product
$E \times B$ points along the cable, and the magnitude of the Poynting vector

$$|S| = \frac{1}{\mu} \left(\frac{U}{r \ln(R_2/R_1)} \right) \left(\frac{\mu I}{2\pi r} \right)$$

has no angle-dependent terms. The power can be calculated as an integral over the cross-section of the dielectric

$$P = \int_{R_1}^{R_2} S \cdot da = \int_{R_1}^{R_2} |S| \cdot 2\pi r dr = UI$$

In numbers, this is a power of 100 kW.

A large ε_r would not change the power. However, the cable would have a larger capacity (a disadvantage with alternating current).

6.2 If the sail is designed in such a way that it completely absorbs the radiation from the sun, then at a power density $S = c^2 g$, the magnitude of the force acting on the surface a is given by

$$F = cga = \frac{1}{c} Sa .$$

Solving for a gives a value for the area of $a = 3 \cdot 10^5$ m^2 if $F = 1$ N. If the sail reflects 100% of the radiation, half the area suffices. In this manner, one gets

$$a_{absorbing} = (548 \text{ m})^2 \quad \text{and}$$
$$a_{reflecting} = (387 \text{ m})^2$$

for the sails.

6.3 Inserting it into the wave equation gives the following result: The field satisfies this equation if $v = 1/\sqrt{\varepsilon\mu}$. Because of

$$\cos x \cos y = \frac{1}{2}[\cos(x + y) + \cos(x - y)]$$

this is also a d'Alembert solution. But it contradicts causality. In the real world, one will not find such a field.

6.4 The solution is based on the law $\nabla \times E = -\partial B/\partial t$. Because of

$$\nabla \times E = (0, 0, -k_x \hat{E}) \cos(\omega t - k_x x)$$

B must always oscillate in z-direction: $B = (0, 0, B)$. And if the time derivative follows the cosine function, the wave must be sinusoidal:

$$B = (0, 0, \hat{B})\sin(\omega t - k_x x)$$
$$\rightarrow \frac{\partial B}{\partial t} = (0, 0, \omega\hat{B})\cos(\omega t - k_x x).$$

Using Maxwell's equation, the result of the rotation can be equated with that of the time derivative: $k_x \hat{E} = \omega \hat{B}$. Also $k = 2\pi/\lambda$ and $\omega = 2\pi f$, leaving $\hat{E} = \lambda f \hat{B} = c\hat{B}$. This result should match ◘ Fig. 6.2.

In general, these two useful relationships are fulfilled:

$$B = \frac{1}{c}e_k \times E \quad \text{and} \quad E = -ce_k \times B,$$

in which $e_k = k/|k|$ is the unit vector in the direction of propagation.

6.5 According to Ampere's law, the magnetic field at the centre of the winding is a ring, and its magnitude is given by $B_Z = \mu NI/R$. An electrical vortex field formed according to the law of induction $\int E dl = -d\Phi_B/dt$. Due to the symmetry of the arrangement, magnetic flux through the circular cross-section area is just as large as it would be in the case of a homogeneous magnetic field B_{center}. Therefore, calling a be the distance from the central ring ($0 < a < r$),

$$\int E dl = E \cdot (2\pi a) = -(\pi a^2) \cdot \frac{dB_{center}}{dt} \rightarrow E = -\frac{1}{2}a\frac{dB_{center}}{dt}.$$

is fulfilled.

For a harmonic progression $B_{centre} = \hat{B}\sin(\omega t)$ one gets $E = -\hat{B}a\omega\cos(\omega t)/2$. The ratio of peak values is then $\hat{E}/\hat{B} = a\omega/2$. It increases with the distance to the central ring. The average energy densities are

$$\frac{w_E}{w_B} = \varepsilon\mu\frac{1}{4}a^2\omega^2$$

and after averaging over a^2 in the range 0 to r, they give a ratio of the total energies

$$\frac{W_E}{W_B} = \frac{1}{12}\varepsilon\mu\omega^2 r^2 = \frac{1}{12}\left(\frac{\omega r}{c}\right)^2,$$

which corresponds almost exactly to the ratio of the field energies in the capacitor if E and B are swapped. Therefore, the conclusion is that mutual field generation can be neglected if the product of the frequency and the largest length of the field line of the generated field is much smaller than the speed of light.

6.6 Since B and the surface elements are perpendicular to each other, the magnetic flux is

$$\Phi_B = \int B \, dA' = l \int B(r') dr' = l \int \frac{\mu I(r')}{2\pi r} dr'.$$

In different parts with different materials, one has

inner conductor $(r < R_1)$ $\qquad \Phi_{B1} = \dfrac{l\mu_1 I}{4\pi}$

insulator $\qquad (R_1 < r < R_2) \quad \Phi_{B2} = \dfrac{l\mu_2 I}{2\pi} \cdot \ln\left(\dfrac{R_2}{R_1}\right)$

outer conductor $(R_2 < r < R_3) \quad \Phi_{B3} = \dfrac{l\mu_3 I}{2\pi} \cdot \left[\dfrac{R_3^2}{R_3^2 - R_2^2} \ln\left(\dfrac{R_3}{R_2}\right) - \dfrac{1}{2}\right].$

The total flow is then $\Phi_B = \Phi_{B1} + \Phi_{B2} + \Phi_{B3}$, the inductance per length is $L/l = \Phi_B/(lI)$.

This solution has some remarkable features: The contribution of the flux in the inner conductor does not depend on its diameter. Often, it is not to be neglected. Because with the same μ for the inner conductor and the insulator the inequality $\Phi_{B2} > \Phi_{B1}$ is only fulfilled from $R_2/R_1 = \sqrt{e} \approx 1.65$ onwards. The contribution of the flux in the outer conductor is the smallest. It is always positive despite the term $-1/2$.

6.7 If the telegraph equations satisfy this symmetry, then in addition to
◻ Eq. (6.40) the equations

$$\underline{u}_2 = \cosh(\underline{g}) \, \underline{u}_1 - \underline{Z}_0 \sinh(\underline{g}) \, \underline{i}_1$$

$$-\underline{i}_2 = \frac{1}{\underline{Z}_0} \sinh(\underline{g}) \, \underline{u}_1 - \cosh(\underline{g}) \, \underline{i}_1$$

must be fulfilled. This can be checked by inserting $\underline{u}_1$ and $\underline{i}_1$ from ● Eq. (6.40) on the right side of the equal sign. For $\underline{u}_2$ one gets the condition

$$\underline{u}_2 = \cosh(\underline{g}) \cdot \left[\cosh(\underline{g})\,\underline{u}_2 + \underline{Z}_0\sinh(\underline{g})\,\underline{i}_2\right] - \underline{Z}_0\sinh(\underline{g}) \cdot \left[\frac{1}{\underline{Z}_0}\sinh(\underline{g})\,\underline{u}_2 + \cosh(\underline{g})\,\underline{i}_2\right]$$

to be always fulfilled. A little rearranging proves the point:

$$\underline{u}_2 = \left[\cosh^2(\underline{g}) - \sinh^2(\underline{g})\right] \cdot \underline{u}_2 + 0 \cdot \underline{i}_2$$

Because $\cosh^2(\underline{g}) - \sinh^2(\underline{g}) = 1$, only $\underline{u}_2 = \underline{u}_2$ remains. Thus, the symmetry is proven for the tensions. The same calculation with the same result can be done for $\underline{i}_2$.

6.8 First, ε_r does not appear in the well-known formula $P = U \cdot I$. Consequently, a reduction in energy transfer through the dielectric is ruled out a priori.

But if the power is to remain unchanged despite the transmission speed reduction, an increased energy density of the electromagnetic wave must compensate the slowing down. And this is precisely what becomes apparent when the Poynting vector (● Eq. (6.31)) is written as the product of velocity $v = 1/\sqrt{\varepsilon\mu}$ and the energy density. The result

$$S = E \cdot \frac{B}{\mu} \rightarrow S = \frac{1}{\sqrt{\varepsilon\mu}}\sqrt{\varepsilon E^2}\sqrt{\frac{B^2}{\mu}}$$

may be interpreted as follows: When the wave penetrates the material, it polarises molecules, which take on energy. Therefore, the electric wave is accompanied by an energy wave passed from molecule to molecule. The material slows down the wave and increases the energy density by the same factor: $\sqrt{\varepsilon}$.

References

1. Juergen Schankenberg: Elektrodynamik, Wiley-VCH Weinheim 2009, ISBN: 978-3-527-40369-1
2. David J. Griffiths, Electrodynamics, 5. Auflage, Cambridge University Press 2023, ISBN 9781009397759

3. Richard P. Feynman, The Feynman Lectures on Physics, New Millennium Edition, Basic Books 2011, ISBN 978-0465023820
4. Patrick J. Sweeney, RFID for dummies, John Wiley and Sons 2005, ISBN: 978-1-118-05447-5
5. ▶ https://www.intel.com/content/www/us/en/products/details/processors/core/i7.html   last seen in 2024
6. Martin Poppe, Pruefungstrainer Elektrotechnik, 4. Auflage, Springer Heidelberg 2022, ISBN 978-3-662-65001-1

6

Supplementary Information

Contents

M. Poppe, *Basic Electrodynamics in 6 Lessons*,
https://doi.org/10.1007/978-3-662-69143-4

Index

GPSR Compliance

*The European Union's (EU) General Product Safety Regulation (GPSR)
is a set of rules that requires consumer products to be safe and our
obligations to ensure this.*

*If you have any concerns about our products, you can contact us on
ProductSafety@springernature.com*

In case Publisher is established outside the EU, the EU authorized
representative is:

Springer Nature Customer Service Center GmbH
Europaplatz 3
69115 Heidelberg, Germany

Batch number: 09254653

Printed by Printforce, the Netherlands